S0-EBT-539

Balancing Agricultural Development and Deforestation in the Brazilian Amazon

Andrea Cattaneo

RESEARCH REPORT 129

INTERNATIONAL FOOD POLICY RESEARCH INSTITUTE
WASHINGTON, D.C.

Printed in the United States of America

International Food Policy Research Institute
2033 K Street, NW
Washington, DC, 20006-1002 U.S.A.
Telephone +1-202-862-5600
www.ifpri.org

Library of Congress Cataloging-in-Publication Data

Cattaneo, Andrea.
Balancing agricultural development and deforestation in the Brazilian Amazon / by Andrea Cattaneo.
p. cm. — (Research report ; 129)
Includes bibliographical references.
ISBN 0-89629-130-8 (alk. paper)
1. Sustainable development—Brazil. 2. Sustainable development—Amazon River Region. 3. Deforestation—Brazil—Econometric models. 4. Deforestation—Amazon River Region—Econometric models. 5. Agriculture and state—Brazil. 6. Agriculture and state—Amazon River Region. 7. Land reform—Brazil. 8. Land reform—Amazon River Region. 9. Agricultural innovations—Brazil—Econometric models. 10. Agricultural innovations—Amazon River Region—Econometric models. 11. Brazil—Economic policy. I. Title. II. Series : Research report (International Food Policy Research Institute) ; 129.
HC188.A5 C334 2002
338.981'07—dc21 2002011967

Contents

Tables

Figures

Foreword

Balancing environmental sustainability and poverty alleviation is particularly challenging in the rainforests of the Brazilian Amazon. Therefore, in line with its mission to identify and analyze policies that help meet the food needs of hungry people without further degrading the natural resource base, IFPRI undertook a three-year research program in the Brazilian Amazon, as part of the CGIAR's initiative on alternatives to slash-and-burn agriculture. This research is also highly relevant for an understanding of long-term climate change linkages.

Although the Brazilian government has recently eliminated policies that inadvertently offered incentives to clear the land, deforestation rates have not decreased. This suggests that there are additional causes of deforestation. From a broad perspective, this research looks at the links between different types of agricultural producers, the logging industry, and the overall institutional setting. It examines these interactions at different levels of geographic aggregation, ranging from survey-based research on small farms in the Western Brazilian Amazon to more aggregate regional and macroeconomic scales relying on secondary data. This report illustrates the economic and environmental effects of macroeconomic and Amazon-wide policies, considers them within a consistent framework, and shows how the Amazon fits into the rest of the Brazilian economy.

To do this, a model was developed to simulate the effects on the Brazilian economy of policy changes, currency devaluation, land tenure regime changes, infrastructure development, and the adoption of new agricultural technologies. The effects of these changes on deforestation and the welfare of farmers and loggers are analyzed in depth. The interesting results, which are at times counterintuitive, shed new light on why slowing deforestation in the Amazon is so difficult, and on the trade-offs between environmental and economic goals.

While this report looks at the Amazon-wide mechanisms at work behind deforestation, a companion IFPRI research report, *Agricultural Intensification by Smallholders in the Western Brazilian Amazon: From Deforestation to Sustainable Land Use,* examines several smallholder settlements on the agricultural frontier. Together, these two reports offer a comprehensive look at an important set of issues around valuable, related resources.

Joachim von Braun
Director General, IFPRI

Acknowledgments

I wish to thank Steve Vosti, Sherman Robinson, and John Boland for the support and training received during the elaboration of this report. I am also grateful to Chantal Line Carpentier, Julie Witcover, and Monica Scatasta for their considerable input and valuable insights, which helped shape this research. Participants in brown-bag seminars in the Trade and Macroeconomics and the Environment and Production Technology divisions at the International Food Policy Research Institute (IFPRI) also provided useful suggestions. I am especially grateful to Peter Hazell for his helpful comments as IFPRI's internal reviewer. To the external reviewers of the manuscript, Tom Tomich and an anonymous referee, I also owe a particular debt of gratitude for a comprehensive evaluation. Their constructive criticism and extensive list of suggestions have vastly improved the quality of the work.

On the Brazilian front, I would like to thank Eustaquio Reis and all the staff at Instituto de Pesquisa Econômica Aplicada in Rio de Janeiro for making this research possible and for their comments. I am also grateful to Marisa Barbosa, Yoshihiko Sugai, Antonio Teixeira-Filho, and the staff at the Brazilian Agricultural Research Corporation (Empresa Brasileira de Pesquisa Agropecuária—Embrapa) in Brasilia for the support they provided, for the crucial data on agricultural technologies, and for the feedback. This research also relied on the support of Embrapa–Acre and Embrapa–Rondônia, and special thanks go to Judson Valentim and Samuel Oliveira. I would also like to thank the participants at the Center for International Forestry Research workshop on "Technological Change in Agriculture and Deforestation," held in Costa Rica in March 1999, who provided useful comments.

For financial support, my gratitude goes to the U.S. Environmental Protection Agency (EPA) for the STAR graduate fellowship that supported me throughout the doctoral program. Substantial financial and logistical support was also given by IFPRI and the Danish development agency (DANIDA) via the Alternatives to Slash-and-Burn Program led by the International Center for Research in Agroforestry. The trips to the field in the Brazilian Amazon would not have been possible without the support of both EPA and IFPRI.

Summary

The Brazilian Amazon, one of the world's largest tropical forests, lost 128,000 square kilometers to deforestation between 1980 and 1995. Agricultural development, logging, and ranching are often identified as the proximate causes. However, the underlying causes of deforestation are rarely discussed in depth.

This report identifies the links between economic growth, poverty alleviation, and natural resource degradation in Brazil. It examines the effects of the following national and regional policies and events: (1) a major devaluation of the Brazilian real (R$); (2) improvements of infrastructure in the Amazon to improve links with the rest of Brazil and bordering countries; (3) modification of land tenure regimes in the Amazon agricultural frontier; (4) adoption of technological change in agriculture both inside and outside the Amazon; and (5) fiscal mechanisms to reduce deforestation rates.

Studying the impact of such phenomena requires an economy-wide view, since the economic activities in other sectors and regions of the Brazilian economy are increasingly linked to those in the Amazon. To this end, IFPRI developed a regionalized computable general equilibrium (CGE) model, which divides Brazil into four regions: the Amazon, Northeast, Center-West, and South/Southeast. In the model, relative product prices, factor availability, transportation costs, and available technology are all assumed to influence land use; biophysical processes as well as decisions of economic agents are assumed to change land cover. Agricultural production activities are broken down by region, sector, and size of operations. A deforestation sector produces arable land used by agricultural producers. Within this framework, land uses (including deforestation), incomes, wage rates, and other aspects of the economy are estimated and differentiated by region.

Looking at the effects of devaluations ranging from 10 to 40 percent, the report finds that under a devaluation of 40 percent, nationally, GDP would decrease, urban poverty would increase, future growth would be undermined, and tradable agricultural goods would expand. In the Amazon itself, a devaluation of 40 percent has these results:

- *Deforestation rates would vary depending on the government's crisis plan.* If the government balances reduction of private consumption, government demand, and investment, deforestation rates would decline by 10 percent in the short run and by 2 percent in the long run. However, government inaction and capital flight would lead to a 6 percent increase in deforestation in the short run and 20 percent in the long run-about 4,000 additional square kilometers per year.
- *Logging would increase* by 16–20 percent depending on government action.
- *The Amazon would fill the domestic demand gap created as other regions move toward tradables.* Following the devaluation, agricultural expansion in the Amazon would

center on production of a variety of annual crops and livestock, as other regions produce more coffee, soy, horticultural goods, and sugar.

The Brazilian government's strategy for Amazonian development, as part of its Avança Brasil (Forward Brazil) plan, includes an ambitious program of infrastructure investments of US$45 billion in 1999–2006. This analysis finds that a resulting 20 percent reduction in transportation costs for all agricultural products from the Amazon would increase deforestation by 15 percent in the short run and by 40 percent in the long run (about 8,000 square kilometers a year). As returns to arable land rise, the incentive to deforest would increase, leading to a 24 percent increase in production by smallholders and a 9 percent increase by large farms. Nationally this would have little effect on welfare, because the increase in production in the Amazon, except for the share that is exported, would replace production from other regions.

Regulating tenure regimes is one of the best options for reducing deforestation in the Amazon. A substantial share of past deforestation occurred at the hands of deforesters who acquired informal land tenure in the process. The Brazilian government is now uncovering fraudulent land claims, reclaiming the land, and moving toward a unified land registry system. Removing the speculative incentive to deforest could reduce the deforestation rate by 23 percent, saving up to 5,500 square kilometers per year.

Agricultural technologies play an important role in determining agricultural development and deforestation. Within the Amazon, the relative profitability and land intensities of different activities, combined with soil productivity and sustainability limits, are factors that affect agricultural producers' incomes and determine, in part, the pressures on forests through the demand for cleared land. The impact of improvements in Amazonian agricultural technologies will depend on which activity is innovated.

- *Livestock technology improvements* appear to yield the greatest returns for all agricultural producers in the Amazon and should improve food security in the region, but deforestation increases dramatically in the long run.
- *Perennial crop technology improvements* could theoretically reduce deforestation rates considerably, but this is unlikely to happen. Small farmers stand to gain the most from such improvements, but they are averse to the risks inherent in perennial crops. Large farmers are unlikely to adopt the new technologies because their gains would be small.
- *Annual crop technology* appears to have little potential. Income gains would be quite small. Before reaching the high land intensity required to reduce deforestation rates, there would almost certainly be a period in which deforestation would increase substantially.

Outside of the Amazon, the agricultural technological change that took place during 1985–95 affected deforestation in drastically different ways. Overall, deforestation rates were 15–35 percent lower than if improvements had not occurred outside the Amazon, largely as a result of innovation in livestock technologies. In fact, improvements in annuals and some perennials alone would have led to a 20 to 27 percent increase in deforestation rates. Regionally, the Northeast and Center-West were the regions to gain in income from technological change. The income distribution gap apparently decreased in the Northeast and increased in the Center-West as a result of technological change outside the Amazon.

To take into account the nonmarket benefits and costs stemming from different land uses, the report considers both taxes and transfer payments. In spite of the link between logging and deforestation, it finds that applying a logging tax in the Amazon would not lead to a decrease in the deforestation rate, but it would negatively affect the logging industry. A deforestation tax would be more effective: a tax of R$50 per hectare deforested would reduce deforestation rates around 9,000 square kilometers a year, with logging only minimally affected. The downside of the deforestation tax is that it would have a substantial negative effect on small farmers in the Amazon.

An alternative scenario would be to subsidize forest conservation. For example, a 30 percent reduction in the deforestation rate could be obtained with a subsidy of R$240 per hectare. From a welfare standpoint, all regions stand to gain from a subsidy of this kind: the Amazon would benefit directly, but the other regions would also gain by taking up the slack in the volume of wood. Market benefits accrued nationwide would exceed the subsidy expenditures. The subsidy, equivalent to R$1.21 per carbon ton of reduced emissions, could be funded internationally if Brazil were allowed compensation for reducing deforestation under carbon trading arrangements with other countries.

CHAPTER 1

Introduction

The primary objective of this research is to identify the links between economic growth, poverty alleviation, and natural resource degradation in Brazil, with particular emphasis on land use and deforestation in the Amazon. This report focuses on the impact of potential macroeconomic policy shifts in Brazil on deforestation and economic welfare, compared with the consequences of technological change in agriculture. The following set of policy questions are applied to Brazil:

- What impact does a macroeconomic shock, such as currency devaluation, have on the movement of the agricultural frontier in the Amazon?
- What will be the economic and environmental impact of forthcoming technological changes in agricultural production inside and outside the Amazon region?
- What are the effects of lower transportation costs resulting from government investments in physical infrastructure in the Amazon?
- What policy mechanisms are most effective in limiting deforestation without hindering economic development? Policies considered are (1) fiscal incentives to account for the forest's value in providing public goods, and (2) the modification of acquisition of property rights in the Amazon agricultural frontier to eliminate inefficient speculative behavior.

The Amazon rainforest covers an area of approximately 5.5 million square kilometers. Sixty percent—3.6 million square kilometers—is located inside Brazil, encompassing nearly 40 percent of the country's territory. In this report the Brazilian Amazon is defined as the North region of Brazil plus northern Mato Grosso and western Maranhão. This specification captures the ecological and agricultural characteristics typical of the tropical forest region. The Amazon so defined, however, still comprises a complex mosaic of forest (72 percent of land area), savanna (15 percent), inundated lowlands (8 percent), and ecological transition areas (5 percent). The savanna areas are important because large areas have been used for mechanized soybean cultivation and for pasture, despite generally poor soils.

The geographic expansion of the Brazilian agricultural frontier has certainly been the most important activity directly involved in the Amazon deforestation process. Agropastoral land uses, particularly cropping and cattle ranching, have been the main cause of deforestation. Timber extraction, charcoal production, mining, and hydroelectric dams have been minor contributors, compared with agriculture, but to the extent that they stimulated agricultural settlements, they have played important causal roles.

Deforestation in the Brazilian Amazon occurs mainly along a band, varying in width between 200 and 600 kilometers. This band stretches from the northeastern state of Maranhão, through Pará and Mato Grosso and includes colonization areas in Rondônia (Figure 1.1). The frontier expansion areas and the government-sponsored colonization areas came into being in the 1960s and 1970s. However, there are areas in the floodplains of the Amazon basin and upland regions in northeastern Pará that were brought into agricultural production in the 19th century. These latter forms of agriculture have adapted over time to the environmental conditions. However, with the onset of roads, floodplain agriculture located along the riparian transportation system lost its attractiveness.

In the state of Pará, upland agriculture is a dynamic and diverse sector of the economy but geographically constrained; therefore, deforestation for agricultural purposes in this report implicitly refers to the frontier expansion areas and the government-sponsored colonization areas.

Since colonial times, the settlement of new frontiers has been undertaken to open access to land and other natural resources. It is assumed that relative product prices, factor availability, and transportation costs are the main economic factors affecting the movement of a frontier. In this publication, the approach taken is the same as that adopted by Findlay (1995), in which frontier movement is described as the process of incorporating a "periphery" into an economic center through "a network of trade, investment, and migration." In the Brazilian context, high transportation costs between the Amazon and the rest of the country—leading to high agricultural input costs and limited interregional trade—characterize the frontier environment. This economic

Figure 1.1 Main agriculture development areas in the Amazon

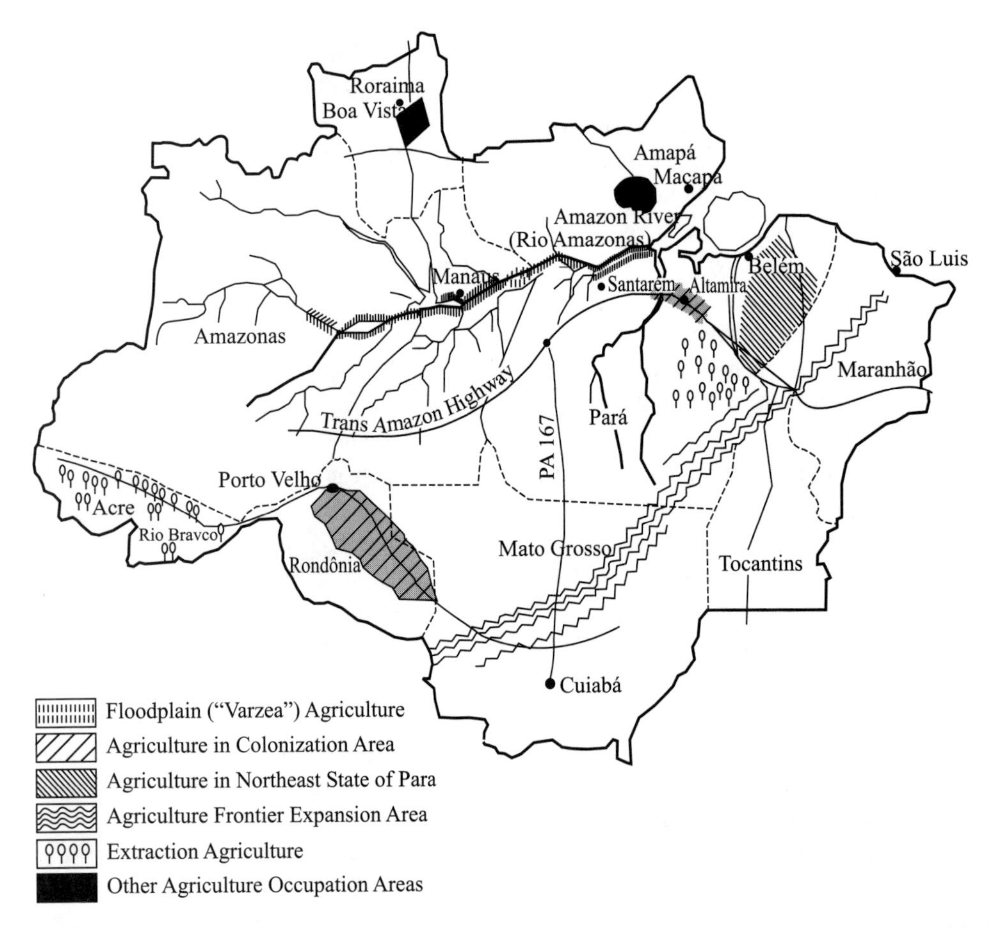

Source: Adapted from Nascimento and Homma 1984. In Serrão and Homma 1993. Reprinted with the permission of Dr. A.K.O. Homma.

intuition is confirmed by the work of Pfaff (1997), who finds that greater distance from markets south of the Amazon leads to less deforestation.

In Brazil, macroeconomic policies, credit and fiscal subsidies to agriculture, and technological change in Brazilian agriculture have all acted as push factors in the migration process. Regional development policies have pulled economic resources through fiscal incentives for agropastoral projects to attract investment, the expansion of the road network to stimulate trade, and colonization programs to facilitate migration (Binswanger 1991). While some of these policies are directed toward reducing poverty, the most harmful ones from an environmental perspective are not driven by equity concerns. The most important example is the fact that agricultural income has been taxed at lower rates than nonagricultural income (a 1.5–6.0 percent tax rate on agricultural income versus a 35–45 percent corporate tax rate in manufacturing and services), thereby converting agriculture into a tax shelter. Small farmers and poor individuals have been negatively affected because the market price of land includes a component capitalizing these tax preferences. This implies that the poor must cut consumption below the imputed value of family labor to pay for the land. Such a policy leads to an increase in deforestation because it creates an incentive for urban investors and corporations to compete for land at the frontiers of settlement as well as in areas of well-established settlement, and because it encourages poor individuals to move to the frontier in search of unclaimed land.

At an intraregional level, several interesting distorting provisions have been reported in the literature, including (1) rules of public land allocation that provide incentives for deforestation because the security of a claim is determined by land clearing (Binswanger 1991); (2) a progressive land tax that encourages the conversion of forest to crop land or pasture (Almeida and Uhl 1995); and (3) a tax credit scheme aimed toward corporate livestock ranches that subsidized inefficient ranches established on cleared forest land (Browder 1988). The fiscal incentives for agricultural production in the Amazon, however, were withdrawn in the late 1980s in response to domestic fiscal concerns plus international criticism of Amazon policy (Lele et al. 2000). With diminished federal support, it was expected that some ranchers, where productivity was low, would abandon their lots, as livestock producers have done in other regions of Brazil. However, profits for pasture systems persisted even with less government support (Faminow and Vosti 1998; Hecht 1993; Mattos and Uhl 1994; Valentim and Vosti forthcoming). While some have argued that intensifying pasture systems could remove pressure to deforest (Mattos and Uhl 1994; Arima and Uhl 1997), they did not always take explicit account of all farm resources (Faminow, Pinho de Sa, and de Magalhães Oliveira 1996) or long-run effects.

Other work at the regional level has emphasized the combined role of expanding road networks and rising agricultural demand in prompting population growth and deforestation (see, for example, Pfaff 1997), while documenting some role for government policy. Using county-level data, Pfaff confirms the importance for deforestation of some of the trends coming out of the policy push to develop the Brazilian Amazon. For instance, development projects were linked to deforestation in the 1970s but not the 1980s (but no robust relationship regarding credit emerged). Closer proximity to markets to the south of the Amazon as well as higher road densities were associated with more deforestation, and early arrivals to a region—not simply higher population densities—had greater impact on the environment. Andersen (1996) similarly found that the importance of federal policy for deforestation faded in the 1980s in the face of local market forces—economic growth, population growth, and locally funded roads.

Schneider (1994) argued that increased road density in already settled areas and fewer roads reaching into new forest areas would help provide sustainable livelihoods for forest inhabitants while protecting further encroachment on the forest. Some studies point to the importance of property rights in Brazilian Amazonian deforestation, including a role for land speculation (Alston, Libecap, and Mueller 1999; Kaimowitz and Angelsen 1998). Still others have found that climatic conditions, principally high precipitation levels, in effect prevent conversion of forest to agriculture (or promote abandonment of that land) even controlling for some market linkages, and that agriculture offers low private returns where practiced (Chomitz and Thomas 2000).

At the local level, an issue open to debate is whether deforestation is primarily carried out by smallholders or by large farm enterprises, and whether the smallholder's goal is to plant crops or install pasture. According to Homma et al. (1998), each of the 600,000 smallholders present in the Brazilian Amazon clears, on average, 2–3 hectares of forest and cultivates it for two to three years. This implies that smallholders clear approximately 600,000 hectares annually. An alternative view holds that commercial ranching has been the largest contributor to the deforestation process. However, as Mahar (1989) points out, some of the deforestation attributed to livestock operations may have been caused by the spread of small-scale agriculture, since land devoted to annual crops is often converted to pasture after a few years when yields decline.

The Brazilian Amazon, with a population of 16 million (61 percent urban), depends to a large extent on local production marketed by both small-scale farmers and large-scale enterprises. While the movement of the agricultural frontier is the major contributor to the deforestation process, the role of agricultural producers in ensuring food security for the region requires a careful analysis of how to reduce deforestation rates without negatively affecting farmers' livelihoods and the regional food supply. Agricultural technologies play an important role in determining agricultural development and deforestation. The relative profitability and land intensities of different activities, combined with soil productivity and sustainability limits, are all factors that affect agricultural producers' incomes and determine, in part, the pressures on forests. The results to be presented in this report compare the magnitude of these effects relative to those of economic processes and policy changes occurring outside the Amazon.

CHAPTER 2

Deforestation in the Brazilian Amazon

Brazil in Context: Comparing Tropical Deforestation Rates Around the World

A comprehensive assessment of the state of the world's forests, released by the Food and Agriculture Organization of the United Nations (FAO), indicates that total forested area continued to decline significantly in the 1990s (FAO 1999). According to FAO's analysis, deforestation is concentrated in the developing world, which lost approximately 62 million hectares between 1990 and 1995 (Table 2.1).[1] The result is an average annual loss in developing countries of 12.5 million hectares. This constitutes a slight decline relative to 1980–90, when annual forest loss in developing countries was estimated at 15.5 million hectares.

Various combinations of agricultural development, logging, and ranching claimed much of the 239,000 square kilometers of forest lost in this South America between 1980 and 1995—the largest loss of forest in the world during those years. Brazil alone lost 128,000 square

Table 2.1 Regions and countries owning major tropical forest stocks and extent of deforestation, 1990–1995 (thousands of hectares)

Region/country	Total forest 1990	Total forest 1995	Total change 1990–95	Annual change	Annual change rate (%)
Africa	538,978	520,237	–18,741	–3,748	–0.7
Congo, Democratic Republic of	112,946	109,245	–3,701	–740	–0.7
Asia	517,505	503,001	–14,504	–2,901	–0.6
Indonesia	115,213	109,791	–5,422	–1,084	–1.0
Central America	84,628	79,443	–5,185	–1,037	–1.3
Mexico	57,927	55,387	–2,540	–508	–0.9
South America	894,466	870,594	–23,872	–4,774	–0.5
Brazil	563,911	551,139	–12,772	–2,554	–0.5
Total developing countries	2,035,577	1,973,275	–62,302	–12,460	–0.6

Source: FAO 1999.

[1]In North America, Europe, and Oceania, reforestation efforts, new forest plantations, and the gradual regrowth and expansion of forest account for an increase of about 6 million hectares of forest cover.

kilometers—more than one-fifth of all tropical forest lost worldwide during that time. Nevertheless, South America maintains vast areas of intact tropical and temperate forest. The northern Amazon Basin and the Guyana Shield house the largest tropical frontier forests anywhere. In fact, the annual deforestation rate for South America, in percentage terms, is lower than the average in the developing world.

Addressing deforestation at a systemic level requires the removal of both market failures and policy failures. While some issues may be addressed at the international level others are best solved at the national level. The data in Table 2.1 concerning standing forests and deforestation highlight the future role that a few countries like Brazil, Indonesia, Mexico, and the Democratic Republic of Congo can play in reducing deforestation rates. These four countries include within their borders approximately 42 percent of standing forest in developing countries and account for 39 percent of all deforestation. Since policy solutions need to be tailored to specific national political and economic environments, it makes sense to focus on those countries that own the largest shares of standing forest and, in particular, on Brazil.[2]

Trends and Geographic Distribution of Deforestation in the Brazilian Amazon

Official estimates of Brazilian deforestation rates are released on an annual basis with a delay of one-to-two years. Forest conversion to agriculture is readily monitored from space using imagery from the Landsat Thematic Mapper (TM) satellites, permitting the development of deforestation maps of large regions at a reasonable cost and speed. As can be seen in Table 2.2, there is considerable variation from year to year and across states. After a substantial decline during 1989–91, the trend in deforestation appears to have spiked sharply in 1995. There is some debate, however, about whether deforestation rates did indeed "spike" in 1995. A possible explanation is that forest losses that took place over the previous two years or so did not register on aerial images due to cloud cover or other complexities of interpretation. If so, what is perceived to be a rapid rise in deforestation in 1995 would instead be a cumulative effect.[3] Another possibility might be that the increase in deforestation during 1993–95 was mainly the result of accidental forest fires (Lele et al. 2000).

In the second half of the 1990s deforestation rates varied between 13,000 square kilometers per year in 1996/97 to 18,200 square kilometers per year in 1999/2000. Although the average deforestation rate in the second half of the decade (approximately 17,000 square kilometers per year) is much lower than the 1994/95 peak (Table 2.2), it is apparently increasing and may return to the 1977–95 historical average of about 19,400 square kilometers per year.

The state-by-state information in Table 2.2 indicates that Pará, Mato Grosso, and

[2]There are countries, for example in Central America, where the need to curb deforestation may be greater because their forest base is smaller and they are deforesting at a faster rate. This is particularly relevant for biodiversity maintenance and local benefits such as hydrologic functions. It is less relevant for greenhouse gas emissions for which tropical deforestation is quite similar no matter where it occurs.

[3]Alves (1999) reports that approximately one-sixth of the area of study was covered by clouds in the first three surveys. In the 1995/96 survey, 30 percent of the area of study was not observed because of clouds. Cloud-covered areas appeared predominantly in Amapá, Roraima, and some areas near the Atlantic Ocean in Maranhão and Pará. The state of Amapá was excluded from the present analysis because of frequent cloud cover over 60 percent or more of its territory.

Table 2.2 Mean rate of gross deforestation by state, 1978–1998 (kilometers²/year)

State	1977/88[a]	1988/89[a]	1989/90[a]	1990/91[a]	1991/92[a]	1992/94[b]	1994/95	1995/96	1996/97	1997/98	1998/99	1999/2000
Acre	620	540	550	380	400	482	1,208	433	358	536	441	547
Amapá	60	130	250	410	36	...	9	...	18	30	...	...
Amazonas	1,510	1,180	520	980	799	370	2,114	1,023	589	670	720	612
Maranhão	2,450	1,420	1,100	670	1,135	372	1,745	1,061	409	1,012	1,230	1,065
Mato Grosso	5,140	5,960	4,020	2,840	4,674	6,220	10,391	6,543	5,271	6,466	6,963	6,369
Pará	6,990	5,750	4,890	3,780	3,787	4,284	7,845	6,135	4,139	5,829	5,111	6,671
Rondônia	2,340	1,430	1,670	1,110	2,265	2,595	4,730	2,432	1,986	2,041	2,358	2,465
Roraima	290	630	150	420	281	240	220	214	184	223	220	253
Tocantins	1,650	730	580	440	409	333	797	320	273	576	216	244
Amazon	21,050	17,770	13,730	11,030	13,786	14,896	29,059	18,161	13,227	17,383	17,259	18,226
Total	42,100	35,540	27,460	22,060	27,572	29,792	58,118	36,322	26,454	34,766	34,518	36,452

Source: INPE 2002.

Note: The leaders indicate a nil or negligible amount.

[a]Mean over the decade

[b]Biennial mean

ERRATA

Page 7, Table 2.2: The numbers in the row across from "Amazon" represent the total for each column. The "Total" row should be deleted in its entirety.

Rondônia have consistently been the states with the largest areas being deforested throughout the 1990s. Where deforestation is occurring is important to this research because, given the aggregate nature of the analysis, it would be ideal to include in the representation of the Brazilian Amazon only those areas that are along the current arc of deforestation or likely to face deforestation pressures in the future.

Alves (2001) presents the geographic distribution of deforestation over the period 1991–96 using a 1/4° grid cell decomposition of the Amazon. These cells are divided by the author into three major "deforestation intensity" categories (high, medium, low), based on the extent of deforestation that occurred during 1991–95 (Figure 2.1). The high intensity cells, for example, are defined as the group of 1/4° grid cells representing 33 percent of total deforestation, that is, the set formed by the cells that represented the most deforested area and amassed 33 percent of total deforestation. The other categories are similarly defined. In Figure 2.1 one can see that a small subset of cells accounted for a large share of deforestation. Alves (2002) reports that approximately 25 percent of the total observed deforestation can be accounted for by just the 3.8 percent of grid cells with the most deforestation while 9.7 percent of the cells accounted for 50 percent of total deforestation. Furthermore, 75 percent of the total observed deforestation is accounted by 19.4 percent of these cells. This shows that the deforestation process tended to be concentrated over an arc extending from Acre and Rondônia in the west, through northern Mato Grosso, into Pará and Maranhão in the east.

Since this report is focused on the deforestation frontier and how it interacts with the rest of the economy, the regional distribution of deforestation presented by Alves was influential in determining the regional disaggregation adopted in the model presented here. In particular, the areas of Mato Grosso, Maranhão, and Tocantins to be included as part of the Amazon were determined by including all cells with high intensity deforestation occurring and the majority of the medium intensity cells (compatibly with the definitions of micro-regions adopted by IBGE).[4] The reason this criterion was adopted is that for modeling purposes, given the aggregate level of analysis, the Amazon should include only economic activities on land that is still forested or along the arc of deforestation. The deforestation frontier shown in Figure 2.1 is an approximate representation of what is considered here to be the border of the Amazon in terms of areas facing deforestation pressures.

Although the deforestation rates reported in Table 2.2 are referred to throughout this report, the exact rate at which the Amazon forest is presently being destroyed is not known. Besides the margin of error associated with ambiguous scenes and cloud cover, the classification by Instituto de Pesquisas Espaciais (INPE) reflects a dichotomy between forest and nonforest that is indeed useful for capturing the main human effects on tropical forests (such as deforestation by ranchers and farmers). But it neglects those forest alterations that reduce tree cover but do not eliminate it, such

[4]The Legal Amazon is defined under Brazilian law as the area comprised by the states of Acre, Amapá, Amazonas, Maranhão (west of the 44° meridian), Mato Grosso, Pará, Rondônia, Roraima and Tocantins. The regional disaggregation adopted here using the intensity of deforestation as the criterion excludes southern Mato Grosso, Eastern Maranhão, and most of Tocantins. The micro-regions in Mato Grosso that were included in our definition of the Amazon are: Alta Floresta, Alto Guaporé, Alto Tele Pires, Arinos, Aripuanã, Colíder, Jauru, Norte Araguaia, Parecis, Sinop, Tangará da Serra. For Maranhão we included: Alto Mearim e Grajaú, Gurupi, Imperatriz, Pindaré, Porto Franco. For Tocantins the included micro-regions are: Araguaína, Bico do Papagaio.

Figure 2.1 Distribution of deforestation in the 1991–1996 period showing deforestation intensity

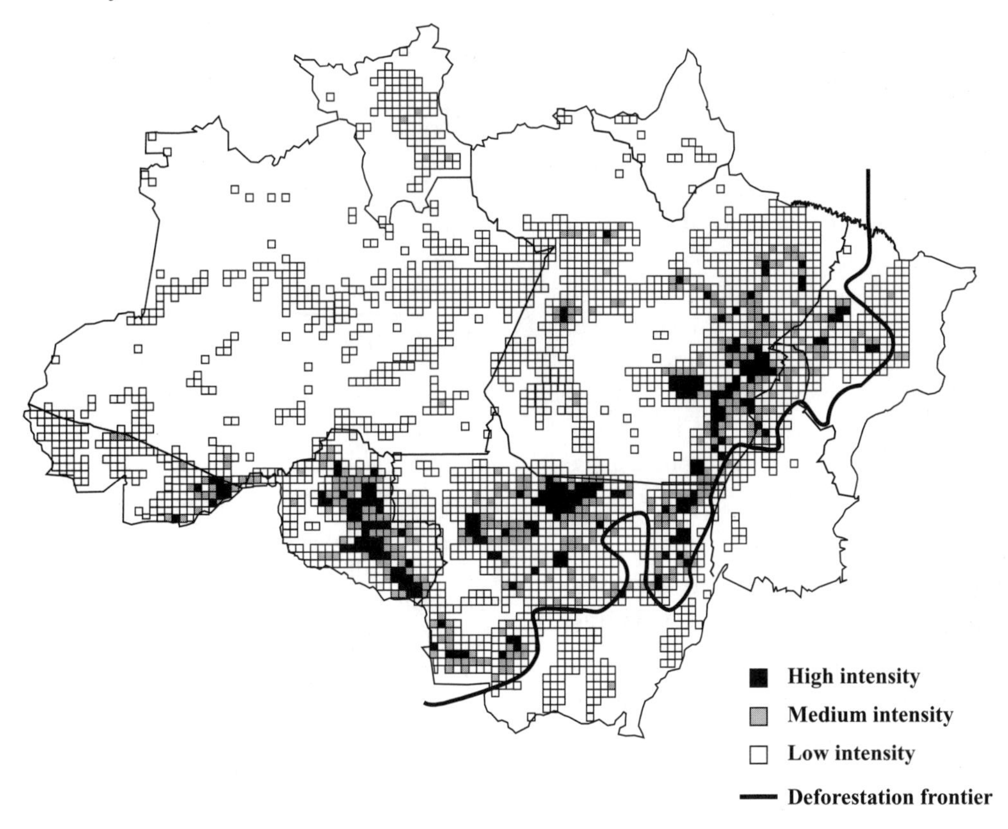

Source: Adapted from Alves 2002.
Notes: The deforestation frontier was defined for this report relying on data in the agricultural census by microregion and on data in Alves 2001 and 2002. Reprinted with permission of Dr. D. S. Alves.

as logging and surface fires in standing forests. The forest openings created by logging and accidental surface fires are visible in Landsat TM images, but they are covered over by regrowing vegetation in one to five years and are easily misclassified without accompanying field data. Although logging and forest surface fires usually do not kill all trees, they severely damage forests. Logging companies in the Amazon kill or damage 10–40 percent of the living biomass of forests by harvesting trees (Verissimo et al. 1992). Based on field surveys of wood mills and forest burning across the Brazilian Amazon, Nepstad et al. (1999) find that logging crews severely damage 10,000–15,000 square kilometers per year of forest that is not included in deforestation

[5]Nepstad et al. (1999), by stating that the area impacted by logging is additional to official estimates of deforestation, implicitly assume that areas deforested for agricultural purposes and those impacted by logging are separate and independent from one another.

mapping programs.[5] While this additional forest area is not explicitly included in the results presented here on the effects of human use, it is taken into consideration when the complementary relationship between logging and deforestation is modeled. This is an important aspect of the modeling effort, given that robust domestic timber demand combined with the exhaustion of forest in Southeast Asia mean logging in the Amazon is likely to grow in the near future. Therefore, even though logging has been determined to be a historically less important factor in Brazilian Amazonian deforestation than agriculture, its role in the region is becoming more significant (Lele et al. 2000; Reis and Margulis 1991).

Table 2.3 Global carbon budget for the 1980s showing major emission sources and uptakes (million metric tons/year)

Source	Emissions
Fossil fuels	5.5 ± 0.5
Tropical deforestation	1.6 ± 1.0
Sink	**Uptake**
Atmospheric buildup	3.3 ± 0.2
Ocean uptake	2.0 ± 0.8
Forest regrowth (Northern Hemisphere)	0.5 ± 0.5
Land sink (by difference)	1.3 ± 1.5

Sources: IPCC 1996.

Greenhouse Gas Emissions from Deforestation in Brazil

Evidence is building in the scientific community that the continued release of greenhouse gases (GHG) threatens to raise the temperature of the earth and disrupt the climates we depend on. Most of the increase in atmospheric carbon dioxide (CO_2) concentrations has come from the use of fossil fuels (coal, oil, and natural gas) for energy, but 20–25 percent of the increase over the last 150 years can be attributed to changes in land use: for example, the clearing of forests and the cultivation of soils for food production. This contribution is confirmed by Table 2.3, which represents the emissions and uptakes (absorption) of carbon during the 1980s, compiled by the Intergovernmental Panel on Climate Change (IPCC). The net release of carbon from changes in land use averaged 1.6±0.7 million metric tons[6] per year, representing 23 percent of the total emissions.

Estimates of the emissions from deforestation in the Brazilian Legal Amazon vary according to the accounting framework adopted: (1) *net committed emissions* refers to the long-term total of emissions and uptakes set in motion by the act of deforestation, and it is calculated only for the area cleared in a given year (for example, the 13.8 x 10^3 kilometers^2 cleared in 1990); (2) *annual balance* refers to the emissions and uptakes in a single year (such as 1990) over the entire landscape (the 415.2 x 10^3 kilometers^2 cleared by 1990). The current best estimate for 1990, according to Fearnside (1999), is 267 x 10^6 tons of carbon for net committed emissions, or alternatively, 353 x 10^6 carbon tons for the annual balance from deforestation (plus an additional 62 x 10 6 carbon tons from logging).[7] The magnitude of these emissions can be appreciated by comparison with global emissions from automobiles: the world's 400 million automobiles emit 550 x 10^6 carbon tons

[6]For the purposes of this report, all tons are metric tons.

[7]Considerable uncertainty still surrounds the overall extent of Brazil's contribution to greenhouse gas emissions. This uncertainty will be reduced when the national inventory, now being compiled by Brazil's Ministry of Science and Technology, is completed. It is following the standardized methodology developed by the IPCC.

annually (Flavin 1989). If one compares Brazil's emissions from land use and cover change with those from fossil fuels (approximately 75 x 10^6 carbon tons), one realizes the importance deforestation has in determining Brazil's greenhouse gas emissions. Therefore, in the future Brazil may stand to gain financial benefits from reducing deforestation if the international community decides it is a viable tool for limiting global warming.

It is now widely accepted that one of the main problems, if not the main problem, for attempts at maintaining forest cover is that it is only rarely a viable financial proposition—while forest exploitation, like one-off logging and deforestation, continue to be highly profitable activities. The global externality associated with greenhouse gas emissions associated with deforestation can be viewed as a case of missing markets for environmental services such as carbon sequestration and biodiversity conservation. International payments, transferring financial resources from consumer nations in recognition of the global public good values of forests, appear to have real potential. Examples of such mechanisms are the Global Environment Fund (GEF), set up in 1991 as a financing mechanism for the International Conventions on Climate Change and Biological Diversity, and the Clean Development Mechanism (CDM) defined under the Kyoto Protocol.[8]

This report considers two categories of corrective actions: one exploits fiscal mechanisms that create disincentives to deforest; the other provides payments (either national or international funds) to compensate producers for the forgone profits associated with reduced emissions. These issues are discussed in greater detail in Chapter 5.

[8]In December 1997 more than 160 nations met in Kyoto, Japan, to negotiate binding limitations on greenhouse gases for the developed nations, pursuant to the objectives of the Framework Convention on Climate Change of 1992. The outcome of the meeting was the Kyoto Protocol, in which the developed nations agreed to limit their greenhouse gas emissions, relative to the levels emitted in 1990.

CHAPTER 3

Modeling Interactions between the Environment and the Economy

This chapter presents a regionalized computable general equilibrium (CGE) model in which Brazil is subdivided into regions compatible with the major administrative subdivisions adopted by the Brazilian government: Amazon, Northeast, Center-West, and South/Southeast. For the Legal Amazon, the following processes are considered: (1) conversion of forests to cleared land (which depends on agents' economic decisions), and (2) transformation of land from cleared land to grassland, and (3) subsequent transformation from grassland to an unproductive state.

The starting point for the regionalized CGE model is a nationwide model developed in 1995–96 as an ongoing collaborative effort between the International Food Policy Research Institute (IFPRI) and the Brazilian National Development Bank (BNDES).

The regional model has two components: a CGE model, which represents the behavior of economic agents, and a land transformation model, which is a simplified representation of biophysical processes affecting land productivity. This chapter begins with a brief survey of the approaches that have been adopted to address deforestation issues using CGE models, followed by a description of the characteristics of the CGE model used for this research and a description of how the biophysical processes are represented.

General Equilibrium Models: From Theory to Practice

In general equilibrium theory, the goal is to formulate a model of simultaneous equilibrium in competitive markets for all commodities that is a precise logical representation of the interaction of consumers and producers. The simplest form of general equilibrium model is the input-output model pioneered by Leontief (1941). In the static input-output model, there is no joint production, only one technique exists for producing each output, and all technologies have constant returns to scale. Input requirements for each unit of output are given by fixed coefficients, and final demand is exogenous. The appeal of this approach is its conceptual simplicity and the tractability afforded by computing equilibrium prices by matrix inversion. The scheme of using matrices to keep track of flows between sectors persists to this day within more complex models of general equilibrium. Isard and Kaniss (1973) give a good account of the uses and shortcomings of the input-output model.

Activity analysis generalizes the production structure by representing it in terms of alternative activities, that is, combinations of inputs and outputs where the ratios between inputs and outputs are fixed in each instance but vary between activities. Joint production is permitted in activity analysis, and there may be more than one activity producing the same output

(Koopmans 1951; Dorfman, Samuelson, and Solow 1958). Within the linear programming environment, prices are assumed exogenous, multiple consumers are not permitted, and the model contains no price distortions. Under these conditions it could be proved that shadow prices coincided with market prices

CGE modeling originated with the work of Johansen (1960). He was the first to introduce a feedback from production levels and endogenous prices to final demand. Johansen solved the general equilibrium model for growth rates by linearizing the model in logarithms and applying matrix inversion techniques. He introduced nonlinear neoclassical substitution possibilities in production and consumption and endogenous determination of market-clearing product and factor prices. The Johansen approach was further developed by Dixon et al. (1982) in their multisectoral ORANI model for the Australian economy. Darwin et al. (1995) and Hertel (1990, 1997) are also in the same tradition.

A technique that is becoming widely adopted is to recast equilibrium problems as *mixed complemetarity problems* (MCP). The MCP is a fundamental problem in optimization that encompasses many of the continuous optimization problems, such as quadratic programming and nonlinear programming, as special cases. It is useful for expressing systems of nonlinear inequalities and equations. A common representation of an MCP has two components: the first represents a set of underlying conditions defined by a system of nonlinear equations, and the second constitutes the complementarity conditions that are only applied to some of the variables and functions. Thc problem can be specified as follows: given a nonlinear function

$$F : \mathbf{R}^n \rightarrow \mathbf{R}^n, \text{ find an } x \in \mathbf{R}^n$$

let I and J be a partition over $\{1, 2, \ldots, n\}$ such that

$$F_I(x) = 0, \quad x_I \text{ free, and}$$

$$F_J(x) \geq 0 \quad \perp \quad x_J \geq 0$$

Where the perpendicular notation "⊥" signifies that, in addition to the stated inequalities, the equation $x^T{}_J F_J(x) = 0$ also holds. For existence and uniqueness of the solution to this problem, see Ferris and Kanzow (1998).

The connection between traditional optimization techniques in economics and this wider problem class was first made by Cottle and Dantzig (1970). A natural connection was also the use of mathematical programming methods in partial equilibrium models pioneered by Samuelson (1949). For a review of papers on the formulation and solution of computable equilibrium problems such as MCP, see Manne 1985; Cottle, Pang, and Stone 1992; Ferris and Pang 1997.

An area that has received wide attention in the field of complementarity problems has been the development of robust and efficient algorithms for solving large-scale applications efficiently. Along with the research in the design of algorithms came the linkage of these algorithms with programming model languages such as the General Algebraic Modeling System (GAMS). The research results to be presented here have been obtained using the PATH solver (available with GAMS), which uses a search method that is a generalization of a line search technique (Dirkse and Ferris 1995).

Modeling Approaches

CGE models have been categorized from analytical through stylized to applied (Robinson 1989). Analytical and stylized numerical models explore the magnitude of the effects of particular causal mechanisms and usually do not provide sufficient detail to analyze and support specific policy recommendations. Applied models consist of a more detailed specification of the institutional side of the country-specific economy under study. Although applied models allow for detailed analysis, there is a danger of concealing the basic causal mechanisms of the model without enhancing its empirical significance, a fact that should be kept in mind when choosing detailed features for an applied model specification (Devarajan, Lewis, and Robinson 1994).

In the domain of the applied models, the detailed nature of CGE models is driven by concerns about policy objectives, external shocks being imposed, and the policy tools being considered to meet the objectives and face the exogenous shocks (Figure 3.1). The combination of these three factors determines the adequate geographic and sectoral aggregations and indicates the appropriate way of representing time. More importantly, the underlying theoretical paradigm will also be affected by these factors.

Although the core of CGE models is neoclassical microeconomic theory, combined with the multisectoral intermediate input links adapted from input-output models, modelers have had to abandon some of the strict neoclassical assumptions in order to meet the imperfections of the actual economies under observation. Instead of perfect competition with perfectly flexible prices and free product and factor mobility, applied CGE models often incorporate

Figure 3.1 Factors affecting the appropriate structure of a CGE model

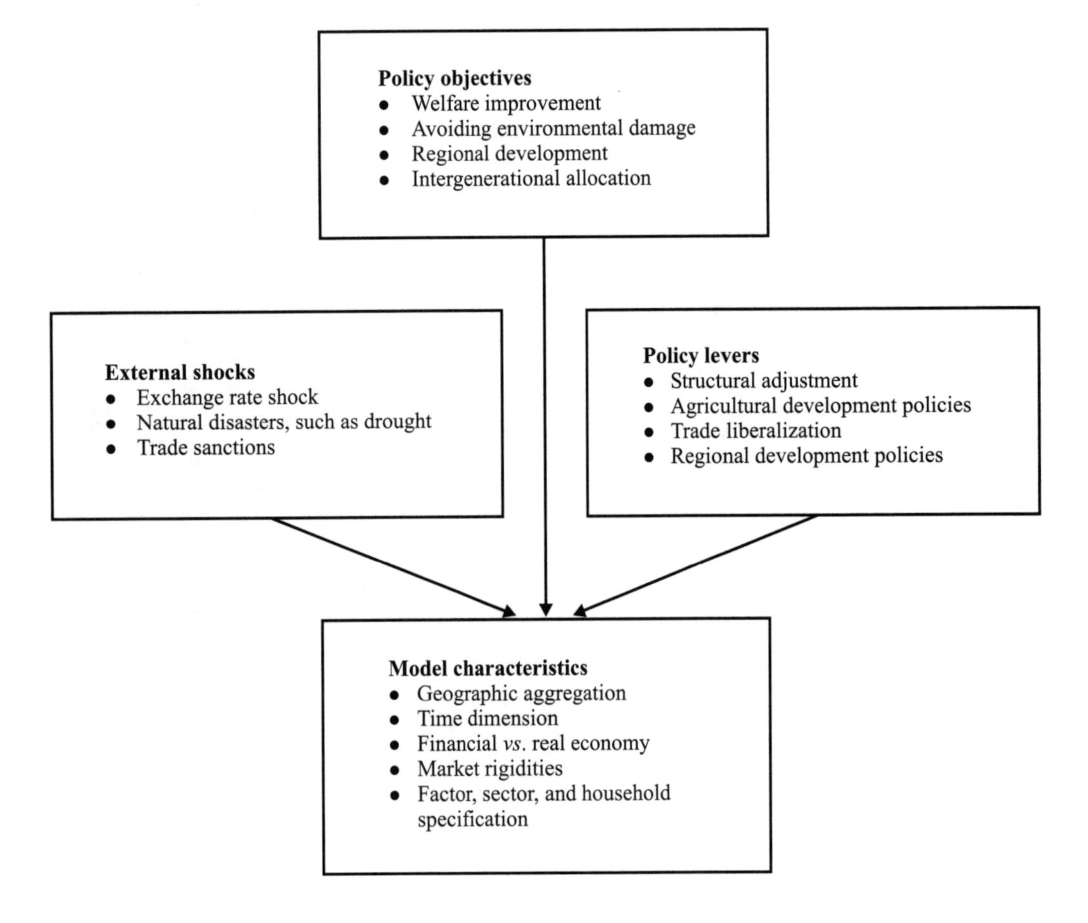

structural rigidities, which seek to capture nonneoclassical behavior, macro imbalances, and institutional rigidities typical of developing economies.[9] The relevant theoretical features that describe macro adjustment, political economy, uncertainty, incomplete markets, and temporary equilibrium are not directly incorporated into the models, but imposed through ad hoc constraints, which are not directly related to the agents' endogenous rational behavior.

Geographic Aggregation

CGE models can be divided into subnational, single-country, and multicountry models. All are open-economy models and incorporate the "rest of the world" as an integral component that permits the consideration of worldwide capital and commodity flows and consequently their influence on the economy under observation. The analytical focus of the study to be carried out determines the geographic aggregation to be applied. Single-country models are used for analyses with a single, national focus. Multicountry models are used to address questions such as global trade liberalization, regional trade agreements, interregional migration, and climate change issues.[10]

Although less common, the focus is sometimes at a subnational level. In such cases one can choose among a spectrum of options for capturing the regionality inside the country. If there are several economically distinguishable regions to be fully represented, a separate CGE model can be constructed for each region connected by flows of factors and commodities as in the multicountry models (Robinson, Hoffman, and Subramanian 1994) .[11] Lofgren and Robinson (1999) present a spatially disaggregated national CGE model that incorporates interregional and national-regional feedbacks to analyze the spatial impacts of economic policies. On the other hand, if regionality is relevant only to a subset of the economic process, such as the presence of a regionally specified activity or factor or both it may be sufficient to maintain a national specification for the model as a whole, while distinguishing the few relevant regional characteristics (Coxhead and Warr 1991; Coxhead and Jayasuriya 1994).

Another reason to model at the subnational level is that the interest is in a natural resource base that is geographically defined. In such cases modeling the single regions, for example, a watershed, may be the appropriate solution (Mukherjee 1996). Isard et al. (1998) present a detailed overview on applied general interregional equilibrium models.

Specification of Time

If the focus of the analysis is comparative statics, the appropriate approach is a single-period model in which all time flow is collapsed into the time before and after an exogenous, unexpected shock. In this case time plays a limited role, agents' expectations are assumed static, only impacts on flows are considered and not impacts on stocks, and the timeframe for adjustment is generally captured by the mobility of factor markets expressed in the model closure.[12] While this approach may appear to be an

[9]These deviations from the Walrasian paradigm and their corresponding methodological problems are criticized in Srinivasan 1982; Bell and Srinivasan 1984; and Shoven and Whalley 1984.

[10]For surveys on this matter refer to Shoven and Whalley 1992; Brown 1992; Goldin, Knudsen, and van der Mensbrugghe 1993; OECD 1990.

[11]In this case care has to be taken because the different regions share a common exchange rate.

[12]Usually in the short-term factors reflect limited intersectoral mobility in the labor markets and none in the capital markets (1 year); in the medium-term labor has full mobility but capital is still fixed (2–4 years); finally, in the long-term both factors are mobile (5–10 years).

oversimplification, it is a useful indicator of the order of magnitude of the impact of a shock or policy measure over an approximate timeframe. At the opposite extreme of the single-period model, there are perfect foresight, intertemporally specified CGE models. This type of model is appropriate when the main focus is the transition path associated with a shock. This interest may arise from a concern with the distribution of income over generations, associated for example with an aging population, or from inefficiencies that could arise from fluctuations in the tax burdens over time. In cases like these, dynamic CGE and other models are best suited to compare the long-term gains of a policy and its short-term costs.

Between the two extremes of static and rational expectations models there is a broad spectrum of options. A deeper treatment of time in CGE models reflects mainly on the stock-flow relationships and the assumptions about agent behavior over time. First, if a model is to be intertemporal, an equation of motion has to be specified to update the factor stocks for labor through population growth, for capital through investment, and for the natural resource base through degradation/regeneration. Second, one must represent the agents' expectations concerning prices and projected incomes. The latter point can be dealt with in a variety of ways ranging from backward-looking expectations (which can be solved recursively) to perfect foresight models (Dixon and Parmenter 1996). The recursive approach is often considered the appropriate choice for capturing the transition path and, in fact, it is often used for forecasting purposes. There are two approaches to macro forecasting in a CGE framework: the first option is to rely on CGE-generated macro implications, and the second is to rely on exogenously supplied macro forecasts, using the CGE model to carry out structural forecasts (Dixon and Parmenter 1996).[13]

The perfect foresight approach is appealing for its model-consistent expectations. Forward-looking models will generally have four distinguishing characteristics. First, consumption is represented as part of life-cycle behavior of consumers. Household behavior is determined by the maximization of an additively separable, time-invariant, intertemporal utility function subject to a lifetime intertemporal budget constraint. Households can be represented as being constituted by *overlapping generations or as infinitely lived agents.*[14] Second, firms are assumed, first, to maximize their market value, which is equal to the present value of their dividend streams, and second, to face imperfect capital mobility due to adjustment costs (q-theory).[15] Third, the government faces an intertemporal budget constraint, and if the government is allowed to run deficits, the debt path is endogenously determined (Pereira 1988; Pereira and Shoven 1988). Finally, the balance of trade and international capital flows have to be specified; not much has been done in this area, and most models assume balanced trade and no capital flows.[16]

---

[13]Forecasting with CGE-generated macro scenarios has not been very successful. When using an external macro forecast, compatibility with the CGE model is ensured by endogenizing variables like total factor productivity and the propensity to save (see Dixon and Parmenter 1996 for more on this matter).

[14]Early work on the overlapping generations dynamic models was done by Auerbach and Kotlikoff (1983); Ballard (1983); Ballard and Goulder (1985); for the infinitely lived agent approach see Bovenberg 1985 and Andersson 1987.

[15]Examples of such firm behavior specifications can be found in Bovenberg 1984; Summers 1985; Goulder and Summers 1987; and Devarajan and Go 1998.

[16]Exceptions are Andersson 1987 and Erlich, Ginsburgh, and Van der Heyden 1987.

Infinite-horizon formulations face severe computational problems when used in applied models. Another drawback of this type of approach is that the baseline to which the simulations will be compared is a balanced growth path (which may or may not occur in reality). Finally, the discount factor, which is generally specified exogenously, will often generate an unrealistic sequence of savings rates (Ginsburg 1994). A good compromise is to build a two-period intertemporal model for a policy measure or shock that takes place during the first period (Erlich, Ginsburgh, and Van der Heyden 1987; Persson 1995).

For an early survey on dynamic CGE models (concentrating on tax policy evaluation), see Pereira and Shoven (1988). In the final part of a book edited by Mercenier and Srinivasan (1994), four contributions by different authors are concerned with modeling intertemporal trade-offs. Azis (1997) compares the impacts of economic reform on rural-urban welfare in a static and a dynamic framework and thereby focuses not only on the economic objectives of the study, but also on the differences of its results with respect to the different methodological approaches. In this vein, Abbink, Braber, and Cohen (1995) demonstrate under what assumptions a simple static CGE model can be extended to a dynamic CGE specification, and they apply both versions simultaneously. Very few applications show explicit interest in and specification of intertemporal aspects of the development process, such as the multisectoral CGE with overlapping generations and intertemporal optimization presented by Keuschnigg and Kohler (1995).[17] Another example is Go (1995), who highlights the intertemporal trade-offs of tariff reforms when examining the sensitivity of investment and growth to external shocks and adjustment policy. Dynamic CGE models are very useful in order to simulate the overall economic development path of an economy. Diao, Yeldan, and Roe (1998) construct a dynamic applied general equilibrium model of a small open economy in order to investigate the transition path and convergence speed of out-of-steady state growth paths in response to trade policy shocks.

Environmental Externalities and Natural Resource Use

Since the 1970s there have been numerous applications of CGE modeling to energy and natural resource issues. Models relating to energy range from those with highly disaggregated specifications of the energy sector, allowing for substitution between energy sources and specifying different demand types, to those focusing more on the rest of the economy, containing a simplified representation of the energy sector.[18] The latter generally focus on the differential impact of a natural-resource boom or crisis on the tradable and nontradable sides of the economy (Benjamin1996; Martin and van Wijnbergen 1986). As an example of the former, Hudson and Jorgenson (1974) constructed an econometric general equilibrium model that captured the interrelationships between energy policies and

[17]Keuschnigg and Kohler (1995) analyze the dynamic effects of trade liberalization in Austria.

[18]Surveys for the disaggregated approach are Devarajan 1988; Bergman 1988; and Bhattacharyya 1996.

economic growth. The authors examined the role of energy taxes in promoting conservation and how to employ the price system to adapt to changes in the availability of energy resources.

The role of taxation to compensate for environmental externalities and its general equilibrium effects are fertile topics for CGE analysis both because the societal costs of such a tax can be estimated through its effect on prices and income (positive analysis), and because optimal taxes may be computed (normative analysis). Jorgenson and Wilcoxen (1990) examine the costs to the economy of emissions regulation and the implications of a carbon tax.[19] For a period there was debate over the so-called "double-dividend" hypothesis, postulating that if the revenue from emission charges is used to reduce the tax on wage income then positive employment effects can result in "second-best" situations with preexisting distortions (Terkla 1984). While this debate has not been resolved, the hypothesis seems to hold only in the short run and under restrictive assumptions (Carraro, Galeotti, and Gallo 1996; Scholz 1998). An interesting development, as the theory of market incentives evolved, was to include markets for tradable emission permits where the equilibrium prices of permits reflect the marginal costs of emission control (Bergman 1991). In reality, the problem with this approach is that a tradable permit program, compared with taxation, has no revenue-raising mechanism to cover the high monitoring costs.[20]

Because of the local and global externalities associated with tropical deforestation, the results presented in the previous paragraphs are important in the context of the research described in this report; however, deforestation occurs mostly on privately owned land. This implies that the economic agent owning the land will view it as an input to production, either agricultural or for timber where externalities are not taken into consideration, or maybe for conservation if externalities are fully internalized. It is therefore important to understand how land as a factor of production is represented in CGE models.

Land is a heterogeneous factor in agricultural production and this poses interesting challenges and possibilities from a modeling standpoint. The productive possibilities of a given hectare of land depend on soil type, drainage, declivity, and climate. These characteristics affect the yield for any specific crop given labor and capital inputs, and therefore determine (along with considerations of the other factors) the most suitable economic activity on a parcel of land. A CGE model focusing on agriculture must manage to capture the constraints on supply response arising from land heterogeneity. Perhaps the simplest method available is to segment the land market along land types that can be put to similar uses. For example, rice and corn can be substituted in production if the land is good, but a producer cannot switch from mediocre pasture to producing rice or corn on that land. This approach implies that activities are either perfectly substitutable or not substitutable at all. A more flexible approach is that adopted by Robidoux et al. (1989) who also differentiate between land types and land uses, but the land types substitute imperfectly in the production of a given crop.[21] In both approaches the land-specific rental rate must be equal across uses. An alternative

[19]See Bhattacharyya (1996) for a survey on the use of CGE for environmental policy analysis.

[20]Revenues can be generated by auctioning off permits, but this one-time inflow will not cover monitoring costs.

[21]The authors of this study on Canada specify constant elasticity of substitution (CES) aggregator functions that combine land types, each of which is used to some degree in each crop.

approach is that adopted by Hertel and Tsigas (1988); they specify a transformation function that takes aggregate farmland as an input and employs it in various uses based on the elasticity of transformation and relative rental rates.

Unlike labor and capital, land is geographically immobile. Regional or climatic differences can be expressed in a number of ways. If farmland is represented as an aggregate input as in Hertel and Tsigas (1988), regionality is difficult to incorporate unless it is embedded in the crop specification. To portray regionality appropriately, land types have to be differentiated along geographic or climatic lines as in Darwin et al. (1995). Land classes are then employed differentially across sectors according to current patterns of production.

This section concludes with an overview of the use of CGE models to analyze issues relating to forestry and deforestation. Following Xie, Vincent, and Panayoutou (1996), CGE models dealing with forest resources can be broadly classified into three groups. The first group consists of applications of standard CGE models that include a forestry sector alongside the other production sectors of the economy (Cruz and Repetto 1992; Coxhead and Jayasuriya 1994; Coxhead and Shively 1995). The second group considers the dynamic nature of forests' reaction to economic processes and resolves the intertemporal forest harvesting problem by modeling a steady state (Dee 1991; Thiele and Wiebelt 1992; Wiebelt 1994; Thiele 1994). The steady-state specification assumes that foresters choose an economically optimal harvest pattern. The limitation of this approach for deforestation in tropical areas such as Brazil is, first, that logging is closer to an extractive process, as opposed to a sustainable, managed forest operation. Second, deforestation is driven mostly by land clearing for agricultural purposes. The third group of models differentiates land uses and types and introduces property rights considerations (Persson and Munasinghe 1995; Persson 1995). They include logging and squatter sectors and therefore markets for logs and cleared land. The model adopted in this paper extends the approach of Persson and Munasinghe (1995) to include land degradation as a feedback mechanism into the deforestation process. A more in-depth review of CGE model applications to deforestation can be found in Kaimowitz and Angelsen (1998).

In their comprehensive review of economic models of deforestation spanning theoretical constructs and scales, Kaimowitz and Angelsen (1998) note some commonality in findings-that ease of access to forest and to long-distance trade paths as well higher agricultural and timber prices or lower rural wages increase deforestation rates. However, problems at each scale of analysis contribute to what Kaimowitz and Angelsen highlight in their review as inconclusive or ambiguous findings about the effects on deforestation of macroeconomic forces, population and migration, changes in productivity and input markets (including land markets and tenure security), and household wealth—or poverty. Since that review, Barbier (2001) has collected papers analyzing deforestation that emphasize economic modeling techniques or that incorporate spatial features and institutional factors (including placement of parks and reserves).

CGE Model Structure: A Primer

In the standard approach to CGE models, one first distinguishes between different agents, such as producers, consumers, and government, and then between goods and factors and the associated markets through which agents interact. The behavioral assumptions of agents are rooted in conventional microeconomic theory: producers maximize profits subject to certain technological constraints (nonincreasing-returns-to-scale production functions) while consumers maximize utility subject to budget constraints, all within the framework of competitive markets. Equilibrium in this type of model is characterized by a set of prices and levels of production such that the market demand equals supply for all commodities. Factors are either fully utilized with flexible market-clearing wages or rent, or alternatively, the wage of a factor has a lower bound below which there is excess supply of that factor. The intersectoral allocation of factors is endogenously determined. The model is specified as a system of nonlinear simultaneous equations. The basic elements of the model can be represented by the circular flow diagram of the economy presented in Figure 3.2. The starting point for the development of this model is a standard CGE model as described in Dervis, de Melo, and Robinson (1982), and the structure of the model draws most directly on Robinson, Kilkenny, and Hanson (1990) and Robinson (1990).

Factor incomes generated by production activities are divided among households in factor-specific shares representing factor ownership. Total household income is used to pay taxes, save, and consume. Government revenue comes from the collection of ad valorem direct taxes and indirect taxes. Government transfers income to households, and expenditure is a fixed share of total absorption. The rest of the world supplies imports and demands export goods. Brazil is treated like a "small country" in the sense that the export demands and import supplies that it faces are infinitely elastic at prevailing prices (with the exception of coffee and sugar).

The macro system constraints (or macro closures) determine the manner in which the accounts for the government, the rest of the world, and savings and investment are brought into balance. On the spending side of the savings-investment balance, nominal

Figure 3.2 CGE structure showing the circular flow of income

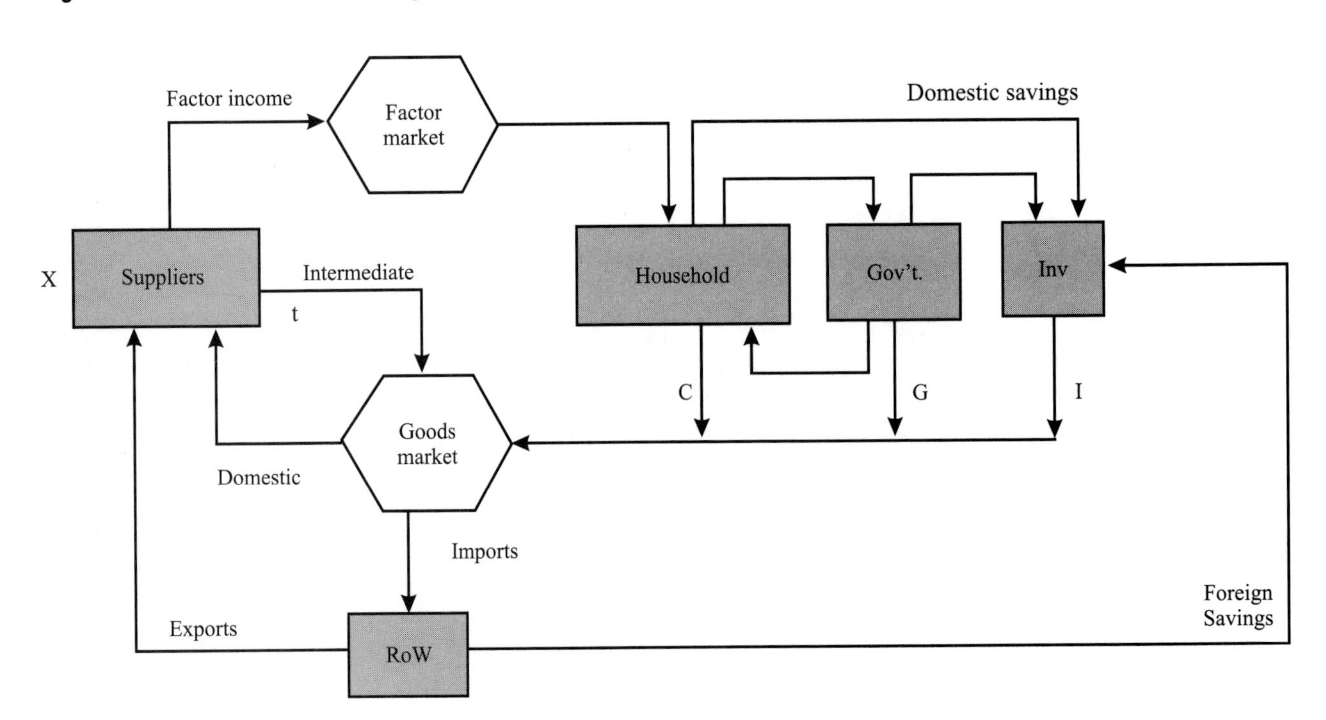

Notes: RoW is rest of world, C is consumption, G is government, and I is investment.

aggregate investment is either a fixed share of total absorption, or it adjusts according to the households' savings rate. On the savings side, if investment is fixed, the average household saving rate adjusts to achieve the level of savings that matches the exogenously specified level of investment. In the government account, total nominal government expenditure is a fixed share of total absorption, and government saving is endogenously determined by the model. Foreign savings is exogenous and the exchange rate adjusts the current account balance.

Model Characteristics

In the modeling approach adopted here, a regionalized CGE model is developed in which Brazil is subdivided into regions compatible with the major administrative subdivisions adopted by the Brazilian government: Amazon, Northeast, Center-West, and South/Southeast.[22] For the Amazon the following processes are considered: (1) conversion of forest to cleared land (depends on agents' economic decisions), (2) transformation of cleared land to grassland, and (3) subsequent transformation from grassland to unproductive states.[23]

The overall model has two components: the CGE model, representing the behavior of economic agents, and the land transformation model, which is a simplified representation of biophysical processes affecting land productivity.

The model allows for two-way trade (cross-hauling) assuming that imports and domestic demand as well as exports and domestic supply are imperfect substitutes (Armington assumption). Producers maximize profits with respect to their nested constant elasticity of substitution (CES) production functions, and households maximize utility with respect to Cobb-Douglas household consumption.[24]

The model is nonfinancial because it does not explicitly include money and asset markets. This choice is based on the assumption that the types of shock considered (changes in the nominal exchange rate, transportation costs, and agricultural technologies) affect most directly the real side of the economy, such as quantities of production and commodities consumed, rather than monetary effects, inflation, and interest rates. While the above hypothesis is somewhat unrealistic in certain situations, the lack of data on the functioning of financial markets necessary to integrate supply and demand variables for money and assets is a limiting factor in modeling financial intermediation of the savings and investment process.[25]

The model is static and solves for a new equilibrium within a single period, given a specified external shock, internal shock, or policy change. The previous section on dynamic CGE models provides some insight into the pros and cons of examining change over time via CGE models used for different analytical purposes. The underlying motivation for choosing a comparative statics approach is that the issues of interest here do not depend on intertemporal optimization by agents, whether it be firms' investment behavior or households' life-cycle saving patterns. The scenarios to be analyzed involve one-time shocks or policy

[22]For the definition of the Amazon region adopted in this report, refer to footnote #4.

[23]The methods adopted could be used to study subsequent regeneration processes through secondary forest growth or planting improved pasture.

[24]The reason for specifying consumption as being Cobb-Douglas is that the income shifts for most of the simulations are sufficiently small that a unitary income elasticity for the components of final demand will not affect the outcome of the simulations for the variables in this report.

[25]See Bourguignon, Branson, and de Melo (1992) for an example of the integration of asset portfolio behavior of macroeconomic models in Tobin's tradition into a CGE model.

measures to which the structure of the economy must adjust in order to return to equilibrium. In terms of expectation models, the shock is a "surprise," requiring adjustments to reestablish the macro balance of the economy.

A complete CGE model also includes a number of closure rules. Closure rules place aggregate constraints on the economic activity simulated in the CGE model. They pertain to how the major macroeconomic accounts (government, trade, labor and capital accounts) adjust to regain equilibrium in response to changes in economic activity. When specifying the model, the system will be overdetermined and one of the constraints of the model must be relaxed to find a solution. Choosing a particular closure rule means precisely deciding which constraint should be dropped. There is no clear-cut theoretical justification for the choice of a particular closure rule except the modeler's general view of an underlying macroeconomic behavior that is assumed exogenous to the CGE model. The closure rules have been shown to have a considerable impact on model structure and the policy conclusions reached (Lysy 1982; Dewatripont and Michel 1987; Robinson 1991). The macroeconomic closure rules of the model and the specification of its factor markets (presented in detail in a later section) will determine the short-, medium-, or long-term character of the model.

The present approach incorporates a number of distinctive model features in order to capture the mechanisms underlying deforestation and agricultural development in a complex setting like Brazil. First, the research is centered on the role of land as a factor of production; therefore, different land classes, with distinct productive possibilities, are specified based on geographic location and vegetative cover. Land in each region is differentiated according to its land type on the basis of cover: (1) forested land, (2) arable land, (3) grassland/pasture, and (4) degraded land.[26] Second, an important characteristic of the marketing process in developing countries with insufficient infrastructure in transport and communication services is the prevalence of high transport and marketing costs. The present approach takes into account this particular characteristic of the economy by incorporating specific marketing margins that are associated with each of the four regions present in the model. This specification allows for detailed analysis of both the economy-wide and regional effects of investment to improve infrastructure. Third, the model incorporates a detailed regional specification of agricultural technologies in the form of multi-output-production functions. The model can therefore take into consideration the ease or difficulty farmers have in shifting production from one crop to another.

The approach is especially useful when considering the impact of technological improvements in agriculture: if an improved technology is not a "substitute" relative to the crops already in production, the impact of technological change will be limited. Fourth, deforestation has been introduced as an explicit economic activity producing cleared land that is demanded by the investment account. For the purpose of this study, this characteristic of the model is of crucial importance because it links agricultural production to the equilibrium demand for deforested land. Demand for deforested land is assumed to be perfectly elastic with the price paid to deforesters determined by the asset value differential between newly

[26]Weed infestation associated with nutrient depletion exhibits a marked threshold effect in soils of the humid tropics, effectively leading to a succession to grassland, whereby farmers' production possibilities are affected from one year to the next. The fact that the effect on farmers of soil degradation is nonmarginal, even though the underlying process is continuous, justifies the assumption that land conditions for agricultural purposes can be expressed by discrete states.

cleared land and forested land, which, in turn, depends on the difference in land rent and on the biophysical degradation affecting the returns to land over time. Fifth, having differentiated land as a factor of production into forested land, arable land, and grassland, each with distinctive productive possibilities, the model keeps track of the stocks of these different land types by factoring in biophysical degradation that transforms arable land into grassland, and grassland into degraded land.

Land Classification

This research is centered on the role of land as a factor of production. Close attention is paid to feedback effects of different environmental states on the economy. Therefore, the principal criteria for identifying land heterogeneity should be the extent to which economic agents' decisions are affected by different environmental states. At this point, the modeler/researcher is faced with an important decision: can the environmental state be described in discrete terms or should it be represented as a continuous process? In other words, does an economic agent react to step-wise or continuous variations in resource quality? This is an important decision from a methodological standpoint because it entails different modeling approaches.

If agents respond to step-wise changes, then it is appropriate to differentiate the resource into a finite number of states, with each of these states having a well-defined role in the economy's production possibilities. This could apply, for example, to qualitative changes in land conditions: a farmer has different options depending on whether the land is forested, cleared, or infested by weeds. In this case three land states can be defined: forest, cleared, and grassland. These would appear as factors of production in different economic activities (for example, forest land in agro-forestry, cleared land in grain cultivation and pasture, and grassland in pasture). Marginal action by the economic agent cannot alter the state of the land. Alternatively, if the agent's productive possibilities are affected in a continuous fashion by changes in resource quality, then it is necessary to incorporate a continuous variable in the production function for each activity, which affects productive possibilities. Where land has no distinct state, but rather its productivity varies along a spectrum based on nutrient levels, then nutrients would be included in the production function. In this case, a marginal action by the economic agent, such as applying fertilizer, would have an impact on production.

As nutrients are depleted in soils of the humid tropics, weeds move in. Weed infestation associated with nutrient depletion marks the threshold of a succession to grassland. Therefore, farmers' production possibilities are affected from one year to the next. The fact that soil degradation is a nonmarginal effect, even though the underlying process is continuous, justifies the assumption that land conditions for agricultural purposes can be expressed by discrete states.

To better describe the approach taken here, it is useful to define some terms and concepts. The differentiation of land into four *land types* on the basis of cover is shown in Figure 3.3. These distinctions are based on the qualitative characteristics that economic agents perceive as making these factors fit for use in distinct economic activities. For example, if land is covered in forest, farmers are able to extract timber or other forest products, but they cannot use the land to plant annuals or perennials, or for pasture, until the land is cleared. Similarly, if the land is cleared and weed infestation has not begun, the land is classified as arable, and can be used for annuals, perennials, or pasture. If the weed infestation has passed a threshold beyond which annuals and perennials are no longer viable, it is classified as grassland and can only be used for pasture. Degraded land is unproductive land and can only be left fallow.

Figure 3.3 Land transformation/conversion flows

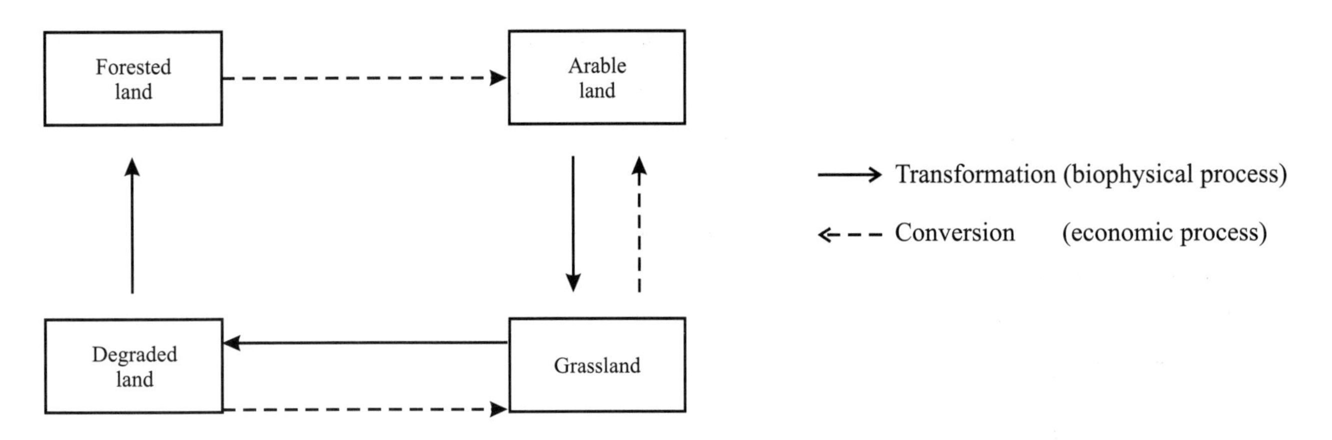

Table 3.1 Mapping of economic activities, commodities produced, and factors used (as adopted in the model)

Activity	Commodities produced	Factors used
Agricultural		
Annuals production	Corn, rice, beans, manioc, sugar, soy, horticultural goods, and other annuals	Arable land, unskilled rural labor, skilled rural labor, and agricultural capital
Perennials production	Coffee, cacao, other perennials	Arable land, unskilled rural labor, skilled rural labor, and agricultural capital
Animal products	Milk, livestock, poultry	Grassland, unskilled rural labor, skilled rural labor, and agricultural capital
Forest products	Nontimber tree products, timber, and deforested land for agricultural purposes	Forest land, unskilled rural labor, skilled rural labor, and agricultural capital
Other agriculture	Other agriculture	Arable land, unskilled rural labor, skilled rural labor, and agricultural capital
Industrial		
Food processing	Food processing	
Mining and oil	Mining and oil	
Industry	Industry	Urban skilled labor, urban unskilled labor, and urban capital - (applies to all sectors)
Construction	Construction	
Trade and transportation	Trade and transportation	
Services	Services	

Land transformations are transitions between land types as a result of physical processes, given certain economic uses. For example, cleared land where rice is cultivated is transformed into grassland.

Land conversion describes a transition between two land types brought about intentionally by economic agents. Usually the agent incurs a conversion cost. In the simulations in this study, farmers cut down trees to plant annuals or perennials or for use as pasture.

Representation of Production and Flow of Goods

The activities considered in the model are presented in Table 3.1, along with the factors employed in production and the commodities being produced by these activities.

Agricultural production is disaggregated by region (Amazon, Center-West, Northeast, Rest of Brazil); by activities (annuals, perennials, animal products, forest

Figure 3.4 Sectoral production technology

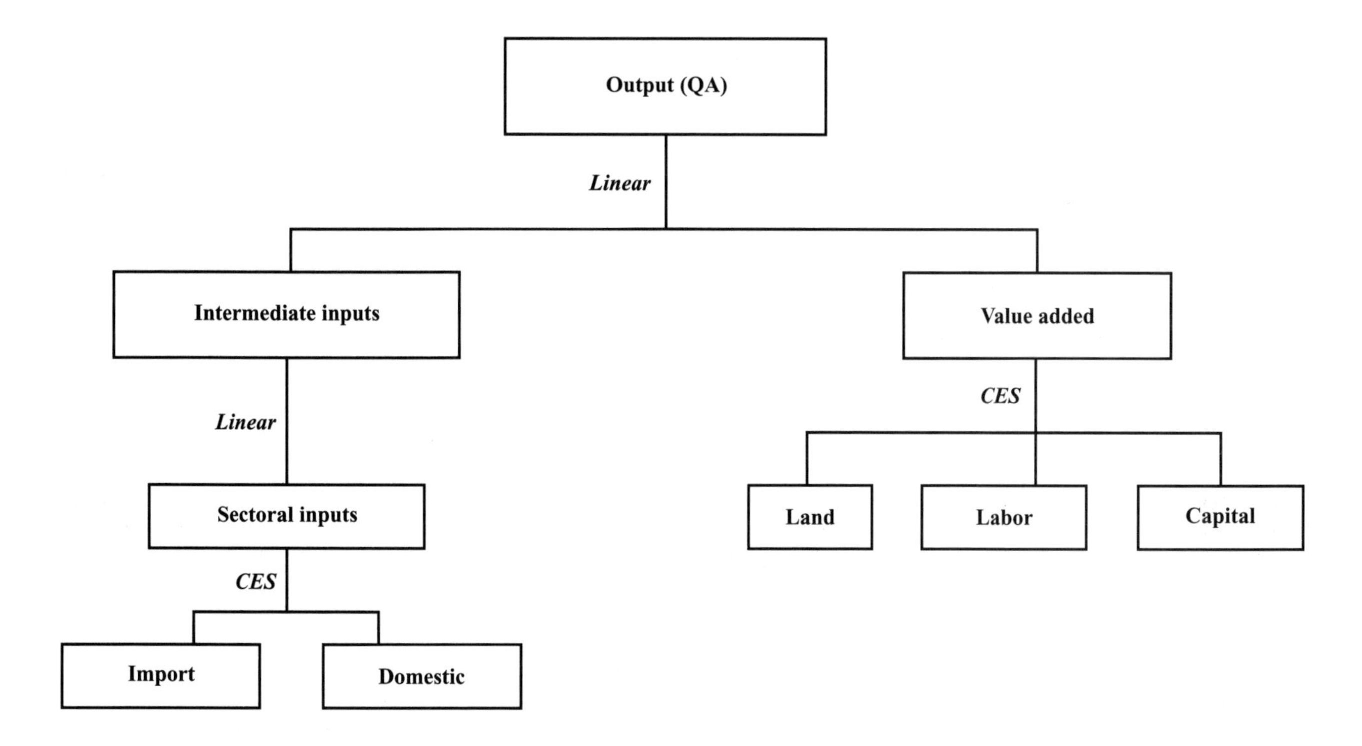

Notes: CES is constant elasticity of substitution.

products, and other agriculture); and by size of operations (smallholder, large farm enterprise). All factors employed by agriculture are region-specific. Producers are assumed to maximize profits given their technology. Agricultural technologies by sector are specified as two-level production functions assuming separability between the two levels. At the lower level, real value added is a CES function of the primary factors of production; output by activity is a fixed coefficient function of real value added and intermediate inputs. The lower level of production technologies is summarized in Figure 3.4.

The Armington assumption is used to capture the choice between imports and domestic output under imperfect substitutability. All domestic demands (including intermediate demands as shown in Figure 3.4) are for the same composite commodity, with the mix between imports and domestic output determined by the assumption that domestic demanders minimize cost subject to imperfect substitutability, captured by a CES aggregation function. This assumption grants the domestic price system a certain degree of independence from import prices and dampens import responses to changes in the producer environment.

The output of the agricultural activity is transformed, at the second level, into commodities, according to a smooth concave transformation frontier described by a translog function obtained as a production-side analogy of the Almost Ideal Demand System (Deaton and Muellbauer 1980). Convexity of the production set was checked according to Hasenkamp (1976). In effect, each agricultural activity produces a number of agricultural commodities ($QA_a \rightarrow QXAC_{a,c}$) (Figure 3.5). For example, a farm producing annuals in the Amazon may have beans, manioc, and rice as

Table 3.2 Production technology: Substitutability between agricultural commodities

Technology	Commodity 1	Commodity 2	Substitutability
Annuals			
production	Corn	Rice, beans	Low
	Corn	Manioc	Low-medium
	Corn	Sugar, soy, horticulture, other annuals	Medium-high
	Rice	Beans	Low
	Rice	Manioc	Low-medium
	Rice	Sugar, soy, horticulture, other annuals	Medium-high
	Beans	Manioc	Low-medium
	Beans	Sugar, soy, horticulture, other annuals	Medium-high
	Manioc	Sugar, soy, horticulture, other annuals	Medium
	Sugar	Soy, horticulture, other annuals	High
	Horticultural goods	Other annuals	Medium-high
Perennials			
production	Coffee	Cacao	High
	Coffee	Other perennials	Medium
	Cacao	Other perennials	Medium-high
Animal			
products	Livestock	Milk	Medium
	Poultry	Livestock, milk	Medium-high
Forest products	Deforested land (agriculture)	Timber	Low-medium
	Deforested land (agriculture)	Nontimber tree products	High
	Nontimber tree products	Timber	High

Notes : The elasticity ranges used are: low = 0.1 to 0.3, low-medium = 0.7 to 0.9, medium = 1.0 to 2.0, medium-high = 2.0 to 4.0, and high = 4.0 to 8.0.

output. This specification allows for the possibility that farmers consider certain agricultural commodities as substitutes and others as complements in the production process. The technology captures both price responsiveness, through own-price elasticities, and technological constraints in transforming agricultural output from one commodity to another through substitution elasticities. Values for these elasticities were obtained by distributing a survey among IFPRI and Embrapa researchers with expert knowledge about the production process in Brazilian agriculture. The results are presented in Table 3.2.

The default option assumes high substitutability in production; at the extreme, it approximates the linear programming farm model approach to production by shifting production to the most profitable crop. If, alternatively, the experts believe that farmers weigh price signals with other factors when making this decision, then substitution elasticities would be lower. Possible factors being considered are (1) relative risk associated with the crops, (2) subsistence requirements, (3) crops requiring similar soil characteristics (substitutable) or different soil characteristics (less substitutable), (4) common practice (habit), and (5) whether intercropping is common for two crops (in this case, at the extreme, there would be very low substitutability).

The general flow from production activities to final commodities is presented in Figure 3.5. The notation for price and quantity variables can be found in the next section on model specification. The diagram starts out at the far left following the contribution of different activities ($QA_1, \ldots, QA_m$) to the production of a single commodity (QX_c) and, moving to the far right, shows

Figure 3.5 Flow of goods from regional producers to the national composite commodity

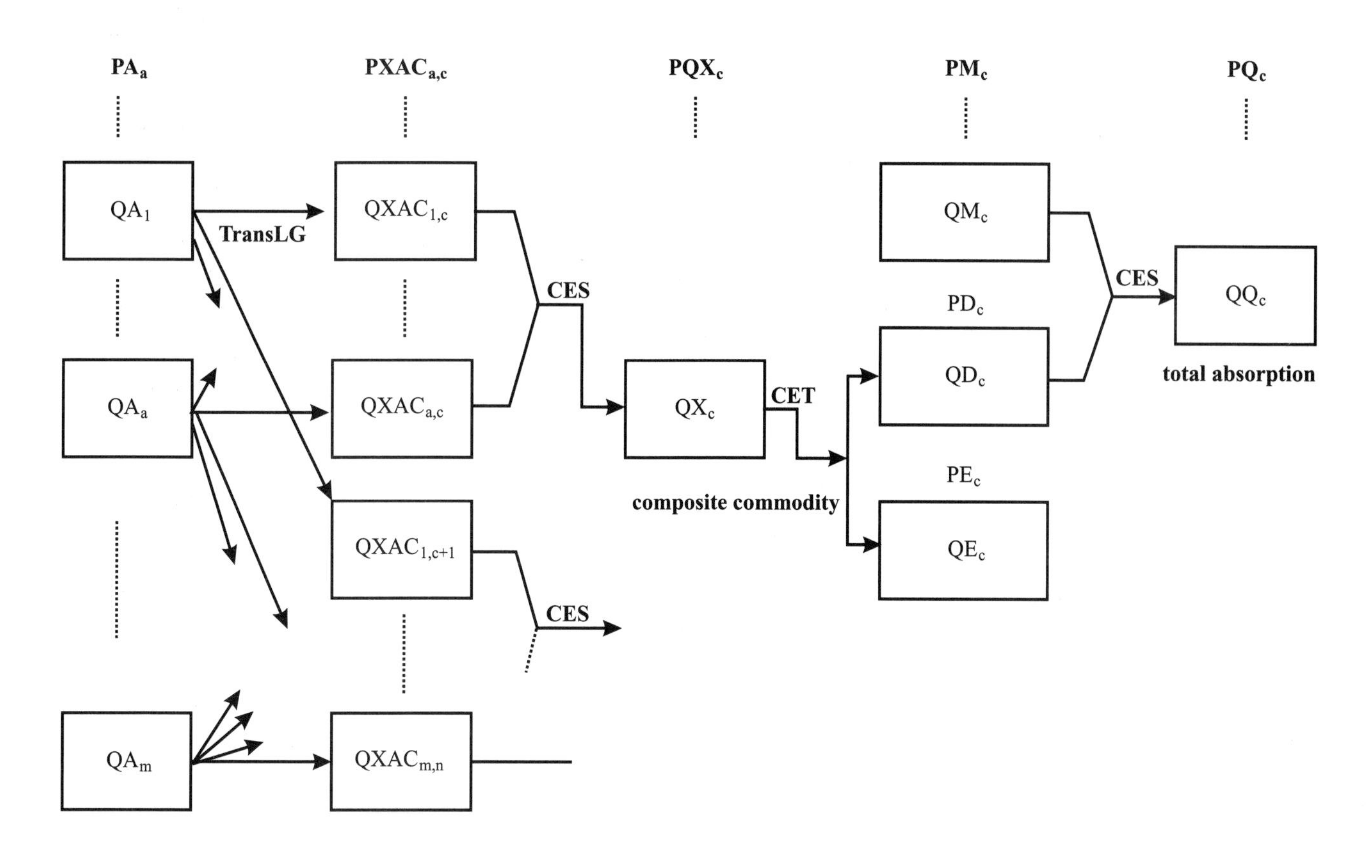

Notes: Prices at top are associated with quantities in boxes as goods are transformed to final product to be consumed or exported. CET is constant elasticity of transformation; CES is constant elasticity of substitution; and TransLG is translog multiple output (agriculture only).

how the domestically produced commodity is affected by the export and import markets.

Outputs are treated similarly to the combination of imports and domestic products. Outputs produced by different regional activities, for a same commodity, are treated as imperfect substitutes in demand in a manner that parallels the treatment of imports and outputs of domestic origin ($QXAC_{a,c} \rightarrow QX_c$, using CES aggregation). The result is that regional activities are allowed a degree of independence from their competitors in other regions of Brazil. This protection arises from the fact that they may produce slightly differentiated goods. Even though only one aggregate national market is considered for each commodity (due to data limitations on interregional flows of commodities), it can also be interpreted that the producers are in reality selling to different market segments (for example, along geographic lines). This allows regions facing higher production or transportation costs in the market for a specific commodity to continue producing.

The allocation of domestic output between exports and domestic sales is determined on the assumption that domestic producers maximize profits subject to imperfect transformability between these two alternatives, expressed by a constant elasticity of transformation (CET) function. This assumption grants the national price system

Table 3.3 Definition of variables, parameters, and indices in the ZimCGE model

Equation	Definition
Sets	
A	Activities
C	Commodities
F	Factors
H	Households
LAND	LAND ($\subset$F)
FCON	Factors involved in conversion ($\subset$ LAND)
FMIG	Interregionally mobile factors ($\subset$F)
Parameters	
α	Share of deforestation occuring on tenured land
$cles_c$	Share of consumption allocated by commodity
$dwt_{f_1 f_2}$	Wage differential threshold for migration to occur between "connected" factor markets
$\overline{FS}_f$	Factor supply in initial equilibrium
$gles_c$	Share of government exp. allocated by commodity
$htax_c$	Household tax rate
i	Discount rate
$itaxac_{af}$	Indirect tax rate
μ_a	Land transformation rate from arable to grassland
μ_g	Land transformation rate from arable to grassland degraded
T	Planning horizon
tm_c	Tariff rate
$wfrat_{f_1 f_2}$	Wage ratio for "connected" factor markets
$yhfc_{f,h}$	Share of factor income to household
$zles_c$	Share of investment allocated by commodity
Variables	
$ABSORB$	Total absorption
CD_c	Final demand for private consumption
$DWG_{f_1 f_2}$	Wage differential between f_1 and f_2
EXR	Exchange rate (R$ per $US)
$FDSC_{f_1 a}$	Factor demand sector
$FSAV$	Net foreign savings
FS_f	Factor supply
GD_c	Final demand for government consumption
$GDTOT$	Total government demand
GR	Government revenue
$HREMIT$	Remittances
ID_c	Final investment demand
$INVABS$	Investment to absorption ratio
$INVEST$	Total investment
MPS_h	Marginal propensity to save
PA_a	Domestic activity price

(continued)

Table 3.3—Continued

Equation	Definition
Sets	
PD_c	Domestic commodity price
PE_c	Domestic price of exports
PM_c	Domestic price of imports
PQ_c	Price of composite good
PWE_c	World price of exports
PWM_c	World price of imports
PX_c	Average output price
$PXAC_{a,c}$	After-tax price of commodity c from activity of a
$PXACP_{a,c}$	Pre-tax price of commodity c from activity a
QA_a	Domestic activity output
QD_c	Domestic sales
QE_c	Exports
$QFCON_{f_1 f_2}$	Factor conversion from factor f_1 to f_2
$QFMIG_f$	Net migration of factor f
QM_c	Imports
QQ_c	Composite goods supply
QX_c	Domestic commodity output
$QXAC_{a,c}$	Domestic output of commodity c from activity a
$SAVING$	Total savings
$UESH_f$	Share of factor f going unemployed
$WF_{f,a}$	Sectoral factor price
$WFAVG_f$	Average factor price
$YFCTR_f$	Factor income
YH_h	Household income
Functional dependencies	
CES	Constant elasticity substitution
CET	Constant elasticity of transformation
$TRANSLOG$	Translogarithmic flexible functional form
$FOC1$	First order condition (FOC) for CES production
$FOC2$	FOC for translog commodity production
$FOC3$	FOC for CET transformation between products for export and domestic markets
$FOC4$	FOC for CES substitution in consumption between import goods and domestically produced goods

Note: See Table 3.4 for the equations for the simplified model and Appendix A, Tables A.1 and A.2 for the full CGE model.

Table 3.4 Model description (simplified version with no intermediate goods)

Equation		Description of equation
Price equations		
1.	$PM_c = PWM_c \cdot (1+tm_c) \cdot EXR$; $PE_c = PWE_c \cdot EXR$	Import price and export prices
2.	$PQ_c = \dfrac{PD_c \cdot QD_c + PM_c \cdot QM_c}{QQ_c}$	Composite commodity prices
3.	$PX_c = \dfrac{PD_c \cdot QD_c + PE_c \cdot QE_c}{QX_c}$	Composite producer prices
4.	$PXAC_{a,c} = PXACP_{a,c} \cdot (1+itaxac_{a,c})$	Commodity prices (including indirect taxes)
5.	$PA_a = TRANSLOG\left(PXACP_{a,c}, Q_a\right)$	Activity prices(multi-output activities)
Quantity equations		
6.	$QA_a = CES(FDSC_{f,a})$	Activity production (CES)
7.	$\dfrac{FDSC_{f,a}}{QA_a} = FOC1(WF_{f,a}, PA_a)$	Demand for primary factors
8.	$QX_c = CES(QXAC_{a,c})$	Commodity demand(CES aggregation)
9.	$\dfrac{QXAC_{a,c}}{QX_c} = FOC2(PX_c, PXAC_{a,c})$	Disaggregated commodity demand
10.	$QA_a = TRANSLOG(QXAC_{a,c})$	Activity production (translog aggregation)
11.	$\dfrac{QXAC_{a,c}}{QA_a} = FOC2(PX_a, PXACP_{a,c})$	Disaggregated multi-commodity production by activity a.
12.	$QX_c = CET(QE_c, QD_c)$	Output transformation (CET) for exporting sectors
13.	$\dfrac{QE_c}{QD_c} = FOC3(PE_c, PD_c)$	Export supply for exports
14.	$QQ_c = CES(QM_c, QD_c)$	Armington assumption:Composite commodity aggregation (CES)
15.	$\dfrac{QM_c}{QD_c} = FOC4(PM_c, PD_c)$	Import demand

(continued)

Table 3.4—Continued

Equation		Description of equation
Income equations		
16.	$YH_h = \sum_{f \in F} \sum_{a \in A} yhfc_{f,h} \cdot WF_{f,a} \cdot FDSC_{f,a} + HREMIT_h$	Household income
17.	$GR = \sum_{h \in H} htax \cdot YH_h + \sum_{a \in A} itaxac_{a,c} \cdot PXACP_a \cdot QXAC_{a,c} + \sum_{c \in C} tm \cdot PWM_c \cdot QM_c \cdot EX$	Government revenue
18.	$SAVING = GOVSAV + FSAV \cdot EXR + \sum_{h \in H} MPS_h \cdot YH_h (1 - htax_h)$	Total savings
Expenditure equations		
19.	$PQ_c \cdot CD_c = \sum_{h \in H} cles_c \cdot (1 - mps_h) \cdot (1 - htax_h) \cdot YH_h$	Household consumption demand
20.	$GD_c = gles_c \cdot GDTOT$	Government consumption demand
21.	$ID_c = zles_c \cdot INVEST$	Fixed investment demand
Factor supply and demand, and migration relationships		
22a.	$FS_{f1} = \overline{FS}_{f1} + \sum_{\substack{a \in A \\ f2 \in f}} \mu_{a,f1,f2} \cdot FDSC_{f_2,a} + \sum_{f2 \in f} QFCON_{f1,f2}$	Factor supply (no migration) Includes factor transformation for physical causes and factor conversion (such as deforestation)
22b.	$FS_f = \overline{FS_f} + QFMIG_f \ for\ f \in FMIG$	Factor supply (with migration)
23.	$QFMIG_{f_1} = \sum_{f_2} (OUTMIG_{f_2,f_1} - OUTMIG_{f_1,f_2})$	Net migration arriving into f1
24.	$WFAVG_f = \sum_{a \in A} WF_{a,f} \cdot FDSC_{a,f} \Big/ \sum_{a \in A} FDSC_a$	Average factor wage
25.	$WFAVG_{f1} = wfrat_{f1,f2} \cdot (1 + DWG_{f_1,f_2}) \cdot WFAVG$	For "connected" factor markets, the wage ratio is constrained
26.	$DWG_{f1,f2} > dwt_{f1} \left[OUTMIG_{f1,f2} > 0 \right]$	Migration occurs when wage differential exceeds threshold
27a.	$\sum_{f \in FMIG} QFMIG_f = 0$	Conservation of total factor supply (for factors that are "connected" through migration)
27b.	$\sum_{f_1,f_2 \in FCON} QFCON_{f_1,f_2} = 0$	Conservation of total factor supply (for factors that are "connected" through conversion)

(continued)

Table 3.4—Continued

	Equation	Description of equation
Factor supply and demand, and migration relationships		
28a.	$FS_f = \sum_{a \in A} FDSC_{a,f}$	Factor market equilibrium (fully employed factors)
28b.	$FS_f > \sum_{a \in A} FDSC_{a,f} \quad \left[WF_f > wf_f^{\min} \right]$	Factor market equilibrium (potentially unemployed factors)
29.	$UESH_f = \left[FS_f - \sum_{a \in A} FDSC_{a,f} \right] \Big/ FS_f$	Share of factor going unemployed (for potentially unemployed factors)
30.	$PX\text{"}def\text{"} = \frac{WFAVG\text{"}ar\text{"}}{i+\mu_a}[\,1-e^{-(i+\mu_a)T}\,] + \frac{WFAVG\text{"}gr\text{"}}{i+\mu_g}[\,1-e^{-(i+\mu_g)T}\,] - \frac{WFAVG\text{"}gr\text{"}}{i+\mu_a+\mu_g}[\,1-e^{-(i+\mu_a+\mu_g)T}\,] - \alpha\frac{WFAVG\text{"}for\text{"}}{i}$	Deforestation demand: price is the expected NPV of returns to land
Macroeconomic closures		
31.	$QQ_c = CD_c + ID_c + GD_c$	Commodity market equilibrium
32.	$\sum_{c \in C} PM_c \cdot QM_c = \sum_{c \in C} PE_c \cdot QE_c + FSAV + \sum_{h \in H} HREMIT_h$	External account balance
33.	$ABSORB = \sum_{c \in C} PQ_c\,(\,CD_c + ID_c + GD_c + DST_c)$	Total absorption
34a. 34b.	$\overline{GOVABS} = \frac{\sum_{c \in C} PQ_c \cdot GD_c}{ABSORB}\ ; \ \overline{INVABS} = \frac{\sum_{c \in C} PQ_c \cdot ID_c}{ABSORB}$	Government consumption and investment demand (fixed share of absorption)
35.	$SAVING = INVEST$	Saving-Investment balance

Source: Compiled by author.

a certain degree of independence from export prices and dampens export responses to changes in the producer environment.

Model Specification

The definitions of the terms used in the model are listed in Table 3.3, and a simplified version of the model used for the simulations is presented in Table 3.4. To highlight the special features of the model, this version ignores intermediate demands, which are treated in a standard way in the full model.

Price Equations

The first set of equations defines prices in the model. On the import and export sides, the model incorporates the "small country" assumption, which states that world prices are exogenous. In the two parts of equation (1), the domestic price of imports and exports is the world price times the exchange rate, with domestic import prices also including a price wedge expressed by the import tax rate (tm_c). The prices of composite commodities (made up of imports and commodities from domestic producers) are defined as a weighted average of domestic and imported commodity prices adjusted for the consumption tax (equation 2). In a parallel manner, for any commodity, the aggregate producer price is a weighted average of domestic sales and export prices (equation 3). The model makes a distinction in equation 4 between price paid for a commodity to an activity based on whether it is at producer prices ($PXACP_{a,c}$) or whether it includes indirect taxes ($PXAC_{a,c}$). The (gross) price paid for any activity (revenue per unit of the activity) is a function of output and commodity prices (equation 5).

Quantity Equations

Equations 6–15 show the quantity equations for commodities and factors that are related to production and foreign trade (the latter only for commodities). Equation 6 defines the CES production function, which, for each activity, determines the relationship between the quantity produced and the use of primary factors. Equation 7 is the demand function for factors, derived from the first-order condition for profit maximization subject to equation 6. Equation 8 defines the demand at the national level for the commodities produced at the regional level. Equation 9 is the first order condition for cost minimization and captures competition between multiple activities (distinguished by their specific technologies) producing a single commodity. Outputs from different activities are imperfect substitutes, an application of the Armington approach (commonly used for international trade) in a domestic setting. In addition to the standard one-to-one mapping between activities and commodities, equation 10 permits multiple outputs for any given activity. More specifically, equation 10 defines aggregate output as a translogarithmic function of output disaggregated by the commodity produced. Equation 11 is a first-order condition derived from cost-minimization subject to equation 10 and a fixed aggregate output demand level. This approach is particularly useful in the context of this project to take into consideration the ease or difficulty farmers have in shifting production from one crop to another.

Equation 12 provides the CET function that transforms domestic output to commodities for exports and domestic sales. Equation 13 is derived from profit maximization subject to equation 12 and a fixed level of domestic output; it defines export supply as a function of relative prices. Equation 14 shows how imports and domestic output sold domestically generate the composite commodities that are demanded by all domestic users. Equation 14 is the Armington function, which is the CES aggregation function for imports and domestic output sold domestically. Equation 15 gives the import demand functions of the relative prices of imports and domestic commodities; it is derived from cost minimization, subject to equation 14 and a fixed level of composite commodity demand. Figure 3.5 summarizes the flow of commodities from production activities to the domestic market and exports. It should be noted that the commodities *QXAC*, *QX*, *QD*, and *QE* are distinct and associated with separate prices (*PXAC*, *PX*, *PD*, and *PE*, respectively). Imports (*QM*) and domestic goods (*QD*) are also distinct from their composite (*QQ*) with separate sectoral prices.

Income Equations

Institutional income flows are extremely simplified in this reduced version. The model institutions are households, government, the savings/investment account, and the rest of the world. Factor income, as a function of factor demand and factor prices, is channeled to the households, and remittances from abroad are also assigned to households (equation 16). Government revenue is defined in equation 17 as the sum of revenue from household taxes, indirect taxes, and import taxes. Total saving, defined in equation 18, is made up of government savings, foreign savings, and household savings.

Expenditure Equations

Domestic final demands are composed of private consumption and investment demand. For each household, consumption is determined by a Cobb-Douglas function, distributing marginal budget share across commodities (equation 19). Similarly, equations 20 and 21 assure that government demand and investment demand are, respectively, allocated across commodities in fixed value shares.

Factor Markets, Migration, and Unemployment

The supply of nonmigrating factors depends on the initial stock, physical transformation, and conversion (equation 22a). Transformation is allowed from arable land to pasture/grassland and from grassland to degraded land. Conversion is allowed from forested land to arable land, and from unemployed arable land to pasture/grassland. In the long run scenarios, interregional mobility of labor and rural capital is assumed. This entails updating factor stocks (equation 22b) based on the balance of in-migration and out-migration for the factor (equation 23). Migration is assumed to rise when there are interregional differences in factor wages, therefore, the average wage of a factor over all activities in which it is employed is defined in equation 24. Keeping in mind that factors are differentiated based on whether they are employed in urban sectors or employed regionally for agriculture, migration is required to maintain the wage ratio between regions in a reasonable range. This is expressed in equation 25 where the wage ratio imposed between two factors is in the neighborhood of a fixed value $wfratf_1,f_2$. The neighborhood of variation for the wage ratio is defined in equation 26, which is specified as a mixed complementarity problem: the wage differential is written as an inequality and linked to the migration variables in the complementary slackness conditions. To allow for interregional differences in the propensity to migrate, the wage differential threshold in the inequality (below which migration does not occur) depends on both the receiving factor (f_1) and the factor providing the migrant flow (f_2). To conclude the migration block, equations 27a and 27b express the conservation of factors, meaning that the net migration and conversion of factors summed over all factors have to balance out to zero.

The equilibrium conditions for factor markets are defined in equations 28a and 28b. It is assumed in the short run that all factors except capital may go unemployed. In the long run only arable land may go unemployed, in which case it is converted to grassland/pasture. Flexible average factor prices perform the task of equilibrating each market. In equation 29, if the lower bound for a factor price becomes binding, a share of the factor will not be employed (*UESHf*). To the extent that it is demanded by different sectors, each factor of production is assumed to be sectorally mobile inside its region.

To conclude the section on factors, the demand for deforestation (producing arable land), expressed by equation 30, will be derived in detail in a later section dedicated explicitly to quantifying the demand for deforested land. In general terms, it expresses the price for arable land as being determined by the returns to agricultural land, which is in turn affected by land degradation. For tenured land, the net returns to deforestation will also depend on the profitability of standing forest (last term in equation 30).

Macroeconomic Closure

Equation 31 is the equilibrium condition for composite commodity markets: supply is set equal to the sum of final demands; flexible composite commodity prices assure that this condition is satisfied. Equation 32 specifies the equilibrium condition for the current account of Brazil's balance of payments. The domestic price index is chosen as numeraire. Foreign savings is fixed (the current account deficit), and the real exchange equilibrates the current account. Absorption is defined in equation 33 as the sum of final demands (investment and government and consumption spending). This definition is drawn upon in equation 34a, which determines the nominal values of investment spending as a fixed share of absorption, and in equation 34b, which similarly determines government spending. Equation 35 defines the final macro closure condition, imposing equality between the values of total savings and total investment.

Demand for Deforested Land

The price for arable land, P_{ar}, is determined by the returns to agricultural land. In an infinite horizon framework, the flow return from an asset divided by the asset price must be equal to the rate of interest in the steady state. What is obtained by going down this path is a perfectly elastic demand for cleared land (which is a reasonable assumption since the investment in newly cleared land is a negligible share of aggregate investment). This implies that the price of arable land for a squatter, *assuming a fixed rental rate*, would be

$$P_{ar} = \int_0^T r_{ar} e^{-it}\, dt = \frac{r_{ar}}{i}[\, 1 - e^{-iT}\,]$$

This expression takes into consideration that an agricultural producer's decision to buy arable land depends on the tenure regime: if the land is subject to insecure property rights, the planning horizon will be finite. A limitation of the expression is that it does not take into account that the rental rate may vary with time due to decreasing or increasing productivity.

For the purpose of this analysis, it is reasonable to assume that arable land is transformed through degradation to grassland, which can be used only for pasture. Let the degradation rate equal μ_a (the indices are dropped to simplify notation) and let r_{gr} equal the rental rate of grassland, then the price for 1 hectare of newly deforested land, if the planning horizon is assumed to be T, is given by the following equations.

Assume $\frac{dA_{ar}}{dt} = -\mu_a A_{ar}$

so that $A_{ar} = A_0\, e^{-\mu t}$ with $A_0 = 1$ (hectare),

then
$$P_{ar} = \int_0^T r_{ar} e^{-it} \cdot e^{-\mu_a t} + r_{gr} e^{-it} \cdot (\, 1 - e^{-\mu_a t}\,)\, dt$$

the solution being:

$$P_{ar} = \frac{r_{gr}}{i}[\, 1 - e^{-iT}\,] + \frac{(\, r_{ar} - r_{gr}\,)}{i}[\, 1 - e^{-(\,i + p\,)T}\,]$$

The interpretation of the last equation is straightforward: the first term represents the value derived from the use of one hectare of land before it degrades to grassland; the second term represents the value derived after conversion to grassland. If there is no land transformation, r_{gr} drops out of the third equation above and the expression simplifies into the first equation. As the degradation rate, μ_a increases the value of a hectare of arable land approaches that of a hectare of grassland.

The above expression, however, does not take into account that the use of grassland for livestock purposes is not agronomically sustainable in many regions of the Brazilian Amazon. To take this additional degradation process into consideration, one must proceed in a manner similar to that adopted to compute the effect of degradation of arable land: if grassland area in livestock use degrades exponentially (after being generated through transformation of arable land) according to

$$\frac{dA_{gr}}{dt} = -\mu_g A_{gr}$$

then the expression for the price of newly arable land becomes

$$P_{ar} = \int_0^T r_{ar} e^{-it} \cdot e^{-\mu t} + r_{gr} e^{-it} \cdot (1 - e^{-\mu_a t}) e^{-\mu_g t} dt$$

$$= \frac{r_{ar}}{i+\mu_a}[1 - e^{-(i+\mu_a)T}] + \left[\frac{r_{gr}}{i+\mu_g}[1 - e^{-(i+\mu_g)T}] - \frac{r_{gr}}{i+\mu_a+\mu_g}[1 - e^{-(i+\mu_a+\mu_g)T}]\right]$$

This is the expression for the price of arable land used in the simplified model (first three terms in equation 30 in Table 3.3). The first term, expressing the value derived before transformation to grassland, has not changed. What has changed, as one would expect, is the value derived from use after conversion to grassland: the limited returns resulting from the degradation of grasslands have now been factored in. As a special case, if μ_g is equal to zero (no grassland degradation) then the equation reverts to the previous case.

The deforesters, being the suppliers of arable land, are faced with this price and the amount of land that will be deforested will depend on P_{ar}, on the returns from forested land (last term in equation 30), and on the squatters' profit-maximizing behavior and technology. The behavior of agents carrying out the land clearing can be differentiated according to whether the forest is an open-access resource or whether property rights governing the use of the forest resource are well defined. For the purpose of this report, it is assumed that the returns to the deforestation activity are based both on acquiring property rights to unclaimed land and on the future returns to agriculture. The net returns to deforestation are different depending on whether land is titled or not; if the land is tenured one must subtract the returns from forested land in the computation. An average return is computed here by taking into consideration that about one-third of agricultural land in the Amazon has been reported to involve fraudulent titles (Brazil, Ministry of Agrarian Development 1999). Therefore, the parameter indicating the share of deforestation occurring on tenured land (α in equation 30) is assumed to equal $^2/_3$. By assuming the planning horizon to be sufficiently long when using arable land, we allow agents to acquire property rights through deforestation.

One last complication, which has not been considered in equation 30, is that arable land can go unemployed and be used as grassland/pasture. If this happens, then the expected returns from agricultural land will be affected, as well as the price paid to deforesters for cleared land. The modified equation 30, as it appears in the full model, taking into consideration the fact that unemployed arable land earns returns equal to those of grassland pasture, is

$$P_{ar} = (1 - UESH_{ar})\left[\frac{r_{ar}}{i+\mu_a}[1 - e^{-(i+\mu_a)T}] + \frac{r_{gr}}{i+\mu_g}[1 - e^{-(i+\mu_g)T}] - \frac{r_{gr}}{i+\mu_a+\mu_g}[1 - e^{-(i+\mu_a+\mu_g)T}]\right] + UESH_{ar}\left[\frac{r_{gr}}{i+\mu_g}[1 - e^{-(i+\mu_g)T}]\right]$$

The equation takes into consideration that the land will be transformed gradually into grassland/pasture (every period a share of land, μ_a, is transformed to grassland). If arable land is fully employed ($UESH_{\text{“ar”}} = 0$), the last term in the equation is zero, and the equation reverts to Equation 30. If arable land is not fully employed, $UESH_{\text{“ar”}} > 0$, the price of newly cleared land is a weighted average of the price of grassland and arable land, and as the share of unemployed land increases, the returns to arable land approximates that of grassland. In the extreme case, where all arable land goes unemployed ($UESH_{\text{“ar”}} = 1$), all newly cleared land will be used as pasture; in this case the price of a hectare of deforested land equals the net present value of a hectare of grassland.

Biophysical Component

This research considers the biophysical processes related to crop sustainability. Among these processes are some that can substantially reduce agricultural productivity, such as soil degradation and weed infestation—problems that usually appear after the first few cropping cycles when a plot is cleared. The biophysical component of the modeling framework affects the equilibrium stocks of the different land types by computing the extent of land transformation given the land uses arising from the simulation. This framework is a first step in linking biophysical changes that occur with a certain land use to the economic incentive for agents to modify existing land use patterns. Including a representation of physical processes in the economic framework is important, because these processes are a major constraining factor for regional development in the Amazon region.

Different productive activities will have different effects on land quality over time. This process belongs to a class of problems that has been studied extensively in the research area known as landscape ecology (Shugart, Crow, and Hett 1973; Horn 1975; Baker 1989; Acevedo, Urban, and Ablan 1995). This research attempts to exploit the analogy between the models developed in landscape ecology, which focus on the succession of ecological states, and the current analysis of the succession of land types given existing land use.

A variety of criteria could be used to distinguish models of land-type change. Perhaps the two most important are the level of aggregation and the use of continuous or discrete mathematics. Models could also be distinguished by the kind of data sources, the method of defining states, and a number of other criteria. A critical research choice that will have to be made early in the research process is the definition of land types. This will likely vary by agro-climatic region. For example, in forest areas and at the forest margins, a possible disaggregation of land types would include pristine forest, arable land, grasslands, and degraded lands, with each type further divided into rich soil and poor soil. In lowland agriculture, the basic division could be between irrigated and rainfed land, with these types further subdivided by soil nutrient status.

The level-of-aggregation criterion refers to the level of detail with which the process leading to changes in land type is modeled. Baker (1989) describes three kinds of models. First are *whole landscape models,* in which the value of a variable in some region is modeled. Second are *distributional landscape models,* in which the distribution of values of a variable in some region is modeled. For example, taking all the land in the region under analysis, one might model the number of hectares falling in each land category (thus losing the differentiation by location). Finally, in the most detailed form are *spatial landscape models,* where the outcome of individual subareas of the landscape and their configuration are modeled. In such spatial models, for example, one could consider the number of hectares in each land category for each farm in the region (ideal if GIS data are available). For this study, a distributional model applied to land types under a given land use is attempted. This choice is necessary because the economic counterpart will consider land use decisions at a regional scale.

Both continuous and discrete mathematics have been used for the time dimension in these models, but there may be little difference in the application of these two approaches. For example, the average response of a stationary Markov process can be obtained by using the corresponding linear constant-coefficient differential equation (Shugart, Crow, and Hett 1973).[27] The matrix approach may still provide an easier framework for modeling changes in variance along with changes in mean. In most cases, empirically based models use estimates of change determined by resampling the landscape at discrete time intervals. The model is in discrete time and the intervals considered are years. The state space is also discrete because a finite number of states in which land can be classified are considered.

Assuming that the process that affects land quality through land use can be described by a land transformation matrix for any farm plot, which can be defined as

$$P = \{p_{fgi}\} \quad f,g = 1,2,\ldots,m$$
$$i = 1,2,\ldots,n$$

where p_{fgi} is the conditional probability that an area of land of type *f* will be transformed into land of type g under activity i between two points in time. Initially, the dependence of the probabilities on the plot's history of land use will be ignored.

The above specification at the plot level is not useful in the context of a model where the unit of analysis is a region like the Brazilian Amazon. To perform the necessary leap in geographic scale, the assumption is made that the regional land stocks by type follow the same transformation pattern.[28] Let L_t be a row vector that specifies total hectares in each land type at time t, then

$$L_{t+1} = L_t P,$$

Figure 3.6 Markov chain representation of biophysical transformation processes

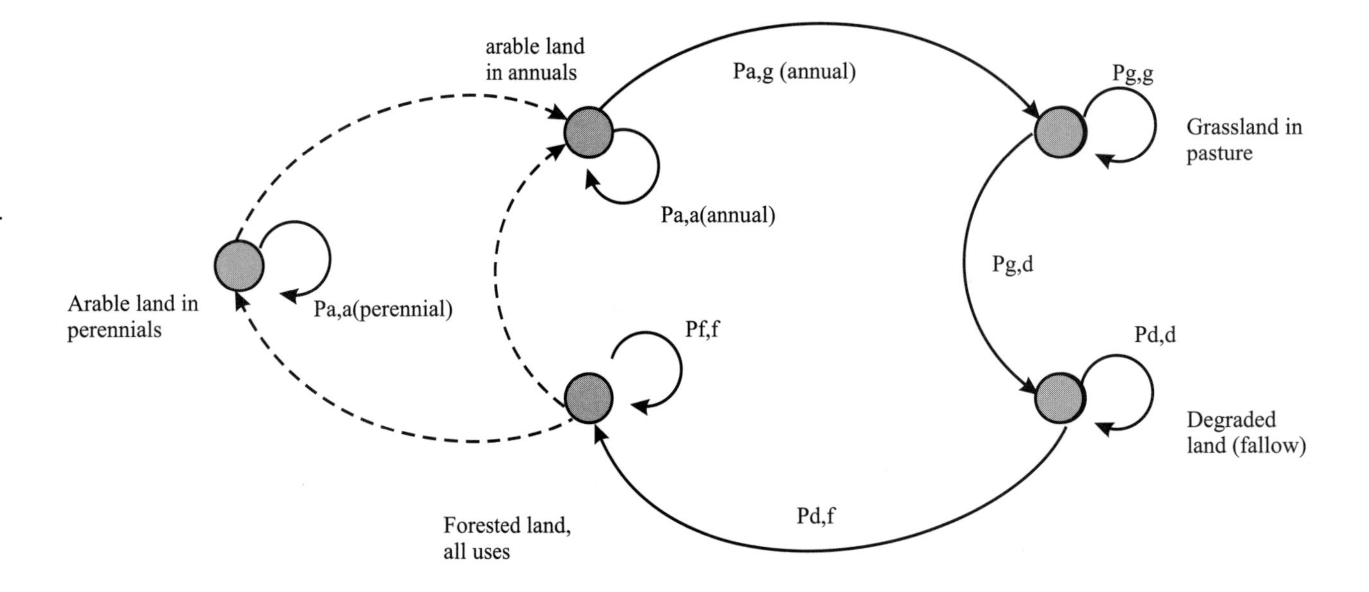

[27]This equivalence in treatment is utilized in this report to the extent that the problem is presented in discrete time for estimation purposes of the transition probabilities, but the representation in the model adopts a continuous time specification (equation 30 in Table 3.3).

[28]Extrapolating from plot level data to Amazon-wide processes can be justified for these biophysical phenomena because, at the simplified level of analysis, the plot differences average out, leaving the important Amazon-specific characteristics that define the problem relative to the rest of Brazil.

where L_{t+1} is a projection at time t+1 of land stocks by type as predicted by the model of the physical transformation process. This is not to be confused with the conversion of land arising from rental differentials in the CGE model. The former expresses a natural physical process (given a fixed land use), while the latter embodies a decision by economic agents to put land to a different use, requiring a physical conversion in land type. With respect to the land types specified in the CGE model, the feasible transformations and conversions are expressed in Figure 3.3.

Technically, the natural transformation process is modeled as a first-order stationary Markov process, with land use entering as an exogenous variable (Baker 1989; Burnham 1973).[29] As shown in Figure 3.6, the system has four biophysical states (forested land, grassland, degraded land, and arable land); the latter, arable land, is divided into two different "exogenous" uses—annual or perennial cultivation—for a total of five states. The probability of remaining in or leaving a particular state is shown in association with the respective arcs ($p_{a,a}$, $p_{a,g}$, $p_{g,g}$, $p_{g,d}$, $p_{d,d}$, $p_{d,f}$, $p_{f,f}$). The probabilities are assumed to be constant for all times into the future. The dashed arcs have no associated probabilities because they are linked to economic decisions that are exogenous to the biophysical component.

The Markov chain approach is reconciled with the static CGE approach by assuming that over an area like the Brazilian Amazon, the probability of transformation can be assumed to correspond to the average transformation. So for example, if $p_{a,a}$ (annuals) in Figure 3.6 is equal to 0.33, this means that at any time 33 percent of the arable land in annuals is being transformed to grassland. Unless there are unexpected shifts in sectoral production from one year to the next, which is improbable, approximating the transformation processes by using the expected value is a valid approach. The probabilities presented here can therefore be used to obtain flows between the stocks of different land types, thereby affecting the equilibrium level of land as a factor of production (equation 22a in Table 3.3). The model solves simultaneously for these flows and for the production pattern, which reflects agents' correct expectations about land degradation processes.

[29]A Markov process is one that describes a stationary stochastic process with discrete, identifiable states, where the future state of the system depends only on the state immediately preceding it.

CHAPTER 4

Model Database

The empirical foundation of a CGE model can be obtained from three sources: (1) econometric methods may be used to estimate parameter values when sufficient time series or cross-section information is available; (2) engineering information may be used to supplement input-output (IO) data to determine technical coefficients of production as well as resource levels; and (3) a comprehensive accounting system may be used to bring all data needed for model simulation into a framework that is consistent with the model specification. For the development of the model for this research all three sources are tapped: estimation methods are used to determine migration parameters and biophysical degradation parameters and engineering information on agricultural technology is incorporated into the database. However, central to the database (and most commonly used in CGE modeling) is the development of a social accounting matrix (SAM). A SAM represents flows of payments between the various actors in the economy. It integrates sectoral, institutional, and national income and product accounts into a unified framework that can be used to analyze the important economic links between factors of production, sectors, and macroeconomic variables.

The structure of a simple SAM is given in Table 4.1. Each cell represents a payment from a column account to a row account. Activities pay for intermediate inputs and factors of production and receive payments for exports and sales to the domestic market. The commodity account buys goods from activities (producers) and the rest of the world (imports) and sells

Table 4.1 A basic national social accounting matrix

	Expenditures					
Receipts	**Activity**	**Commodity**	**Factors**	**Rest of world**	**Institutions**	**Totals**
Activity	...	Domestic sales	...	Exports	...	Total value of production
Commodity	Intermediate inputs	...	...	...	Final demand	Total demand
Factors	Value added	...	...	...	...	Total value added
Rest of world	...	Imports	...	...	...	Foreign exchange outflow
Institutions	...	...	Factor income	Trade balance	...	Gross national income
Totals	**Total costs**	**Total absorption**	**Total factor income**	**Foreign exchange inflow**	**Total absorption**	...

commodities to activities (intermediate inputs) and final demanders (households, government, and investment). In this simple SAM, sectoral specification, interregional flows, tariffs, indirect taxes, and subsidies are left out.

The matrix of column coefficients from such a SAM provides raw material for much economic analysis and modeling. The intermediate-input coefficients correspond to Leontief input-output coefficients. Column coefficients provide the starting point for estimating parameters of nonlinear, neoclassical production functions, factor-demand functions, and household expenditure functions. Given that so many of the model parameters depend on the flows in the SAM, it is necessary to understand thoroughly the data framework.

The micro SAM developed in this study is in many ways standard. In constructing the SAM, the general approach presented by Pyatt and Round (1985) has been followed. This SAM incorporates disaggregated agricultural flows in its description of the Brazilian economy, and, in so doing, fills a gap in Brazilian IO data. Agricultural production is differentiated by region. The study uses the 1995 IO table for Brazil, Embrapa data on agricultural production technologies, and incorporates agricultural census data for 1995–96 (IBGE 1995, 1998a).

The SAM for this research starts with a large amount of detail on the production side (205 activities). With 23 agricultural categories, differentiated by small farms or large estate production for each of the four regions adopted in the model (184 activities), a deforestation activity, and 5 food-processing activities, the agricultural sector is extremely well represented. The population of Brazil is 24 percent rural, with regional peaks of 40 percent in both the Amazon and the Northeast (IBGE 1998b); the majority of the rural people depend upon agriculture for their livelihoods. Consequently, detail in the agricultural sector is highly desirable for analyzing poverty alleviation, development strategy, and deforestation.

Trade and transport margins are also important for commodities in an interregional context. Because distances are large and transaction costs high, the difference between the market price and the price at the farm or factory gate can be significant in certain regions of Brazil. Domestic marketing margins are explicitly broken out here for each regional activity in the micro SAM.

Several household types are taken into consideration so that welfare implications on different income groups of different policy scenarios can be analyzed. In addition to the producers and the households, the other actors in the economy are the government, investors, and foreign demanders or suppliers.

The main central government agency involved in the collection, analysis, and dissemination of such information is the Brazilian statistical institute, Instituto Brasileiro de Geografia e Estatistica (IBGE), which reports to the Ministry of Science and Technology and consists of a number of national directorates responsible for data collection. Current IBGE national accounts are based primarily on the following sources.

- The 1991 demographic survey, which provides IBGE with information regarding total population by region and the distribution of employment between activities
- Household surveys (1987 and 1996) taken for nine major urban areas in Brazil. Because resources are limited and population density low outside the cities, neither survey ventured deeply into rural areas. It is a considerable drawback that the standard of living and consumption patterns of rural households are only partially represented within the sampling frame of the major cities surveys.

- External trade data based on customs declarations
- Available government accounts. The informational content of these data are limited by the fact that recurrent expenditures and investment expenditures are not presented separately. The value added in the government sector consists of compensation paid to employees.
- Agricultural production surveys. The Agriculture Division of IBGE produces estimates of total production of basic food crops and marketed production of other important agricultural commodities.
- Industrial data collection. Industrial data are available from a variety of sources including labor force and salary surveys, industrial production surveys, surveys of the construction industry, intermediate consumption and inventory measurements, and a business enterprise survey. These data are collected at regular intervals (some monthly, some by trimester, and some annually).
- Trade margins are calculated as the difference between the price on goods sold and the cost of purchasing the goods (by the wholesaler or retailer). This information consistently indicates a high trade margin.

The data are compiled in accordance with the United Nations System of National Accounts (SNA) to as great a degree as possible. Useful information from a variety of different institutions, which will be referred to in more detail in subsequent sections, are also drawn together.

IBGE's estimation of gross domestic product (GDP) is based on the commodity-flow approach. This relies on the supply and demand for 100 product groups. The breakdown of total demand into intermediate demand, final demand, and capital formation is based on estimated technical coefficients. While potentially inaccurate, the technical coefficient approach is necessary since actual data are not available (IBGE 1997b).

Despite the shortcomings associated with piecing together data from different sources, the IBGE national accounts are the best set of information available. It is true that much desirable information is either unknown or of uncertain quality. It is also true that much is known about basic production structure (in agriculture as well as industry), consumer habits, government spending and revenue, structure of imports and exports, and financial flows to Brazil. In developing the 1995 Brazil SAM for this study, efforts were made to maintain as close a correspondence to IBGE national accounts as possible.

The year 1995 was chosen as the benchmark because, with the release of the 1995/96 Agricultural Census, it is the most recent year for which comprehensive and reliable data are available (the previous agricultural census was for 1985). And 1995 can certainly be considered a more normal year than any year in the previous decade. In 1995, a comprehensive stabilization plan (*Plano Real*) helped check inflation in Brazil, and no exogenous shocks such as drought hit the economy. Finally, by 1995, the process of removing distorting government policies was under way and privatization of state enterprises had begun in earnest. The tangible differences in the economy between 1994 and 1995 and the superior quality of some statistics, provide ample reason for using 1995.

A Macroeconomic Social Accounting Matrix (MACSAM)

MACSAM entries are in the form of macroeconomic aggregates. In a SAM, rows track receipts, while columns track expenditures. Hence, row sums represent total receipts and column sums represent total payments by a given account or institution. In the tradition of double-entry accounting, row sums must equal column sums. A complete discussion of the economic relationships embodied in a SAM can be found in Pyatt and Round (1985).

Table 4.2 lists data sources and a brief description of how the value of all the relevant entries (cells) in the macroeconomic SAM were found. With the original values, the MACSAM for Brazil balances exactly (row sums equal column sums) by establishing certain cells as residuals (Table 4.3).

A Disaggregated Social Accounting Matrix for 1995

To allow for more detailed policy experiments and to establish the basis for a microeconomic CGE, the MACSAM developed in the previous subsection must be disaggregated. The procedure applied here strives to develop a balanced micro SAM, BRASAM, while maintaining as close a

Table 4.2 Data sources for macroeconomic social accounting matrix

Row	Column	Source	Description
Activities	Commodities	NA Table 2, Section CI	Sales of marketed production at producer prices calculated from gross value of production
Commodities	Activities	NA Table 2,Section OBS	Intermediate consumption
Commodities	Households	NA Table 2,Section DF	Marketed consumption by households
Commodities	Government	NA Table 2,Section DF	Total government expenditure including salaries
Commodities	Rest of world	NA Table 2	Total export revenue (FOB) (includes export taxes)
Commodities	Saving andinvestment	NA Table 2,Section DF	Total investment (includes inventory changes)
Labor	Activities	NA Table 2,Section VA	Labor component of value added at factor cost
Capital	Activities	NA Table 2,Section VA	Capital component of value added at factor cost
Enterprises	Capital	Implied	Gross profits to formal enterprises
Households	Labor	Residual	Private and public sector wages
Households	Enterprises	Residual	Distributed profits. Equals income of formal enterprises less enterprise taxes, retained earnings, and depreciation
Households	Government	IMF/IFS	Government transfers to private households. Social security payments plus interest payments to domestic creditors
Households	Rest of world	IMF/IFS	Foreign remittances to households. Net remittances of workers
Government	Commodities	NA Table 1,Imports Section	Tariffs paid on imports
Government	Enterprises	IMF International Financial Statistics	Enterprise taxes
Government	Households	Government accounts	Income taxes
Government	Indirect taxes	Implied	Government receipts of indirect tax revenue, equal to output taxes plus import tariffs less export subsidies
Indirect taxes	Activities	NA Tables 1 and 2	Output taxes. Comprised of per unit output price taxes and interstate production tax
Saving and investment	Enterprises	Estimated	Retained earnings plus depreciation
Saving and investment	Households	Estimated	Private savings
Saving and investment	Government	Implied	Government savings. Government expenditure less government receipts. The cell adjusts to balance government consumption row and column totals
Saving and investment	Rest of world	Residual	Net capital inflow. This cell ensures balance between foreign exchange availability and imports of goods and nonfactor services
Rest of world	Commodities	NA Table 1,Imports Section	Imports

Source: IMF 2000, IBGE 1997b.
Note: NA is national accounts (from IBGE); IFS is international financial statistics (from IMF).

Table 4.3 Macroeconomic social accounting matrix for Brazil, 1995 (R$ current billion)

	Expenditures											
Receipts	Activities	Commodities	Labor	Capital	Firms	Households	Government	Direct taxes	Indirect taxes	Saving and investment	World	Total
Activities		1,267.96										1,267.96
Commodities	615.36					429.75	110.48			126.64	46.31	1,328.55
Labor	339.32											339.32
Capital	222.47											222.47
Firms				222.47								222.47
Households			339.32		80.51		95.66				3.50	518.98
Government		5.54						79.78	90.82			176.15
Direct taxes					27.92	51.86						79.78
Indirect taxes	90.82											90.82
Saving and investment					102.98	37.37	-30.00				16.29	126.64
World		55.05			11.05							66.10
Total	**1,267.96**	**1,328.55**	**339.32**	**222.47**	**222.47**	**518.98**	**176.15**	**79.78**	**90.82**	**126.64**	**66.10**	

correspondence as possible to the national accounts data. To achieve this goal, the procedure is divided into three steps, involving, first, the construction of a raw, unbalanced BRASAM at the greatest level of disaggregation allowed by the available data. Activities are disaggregated into 205 subgroups, commodities into 44 subgroups, and factors into 23 categories according to Table 4.2. Second, the unbalanced BRASAM is aggregated to the desired sectoral specification for the model. Third, the aggregated version of the SAM is balanced.

The original sources used to construct the SAM are the 1995 IO table for Brazil (IBGE 1997a) and national accounts (IBGE 1997b). These sources are integrated with the agricultural census data for 1995-96 (IBGE 1998a) to yield a regionalized representation of agricultural activities. Household data are obtained from the national accounts and the household income and expenditure surveys (IBGE 1997c and 1997d). For technology coefficients in agriculture, the SAM relies on information from an Embrapa database with detailed regional specification of technologies by crop type.

The cell entries in the raw BRASAM present a picture of the economy in 1995, which is taken as prior information. However, as a result of missing information, data inaccuracies, incompatibilities between micro and aggregate data, and accounting discrepancies, the row and column sums of the raw BRASAM do not balance, even if the macro totals implied by the raw BRASAM satisfy the values in MACSAM.

A general point to be kept in mind throughout this section is that the totals in the 205 activity columns, which are evidently of critical importance since only 28 sectors are identified in the national accounts, were established as follows for the raw BRASAM:

(1) Down the columns, total costs of production (including factor use and output taxes), that is, total payments, are directly available from the national accounts for all activities at a *national level* of aggregation, except for 12 agricultural activities that are not present even at the national level. Data for the 20 activities where direct mapping is possible are therefore entered into the raw BRASAM in unchanged form.
(2) For the 11 agricultural activities available at the national level, total national sales figures (disaggregated totals in the activity rows) are available in the national accounts. These totals are subdivided into 44 regional activities through the use of the agricultural census regional production surveys for the Amazon, Northeast, Center-West, and South/Southeast Brazil. The column coefficients, representing the regional technology, are obtained from a weighted average of coefficients provided by Embrapa at a microregional level.
(3) For the 12 agricultural activities not available even at the national level, regional production surveys are again used and the total for each of these activities is subtracted from the "other agriculture" activity total.

The primary source of discrepancies between row and column sums in the raw BRASAM developed in this section stem from the elements in the activity columns, which contain information on IO relationships, factor use, and output taxes.

An intermediate consumption or IO matrix shows activities in the columns and commodities in the rows. Each activity purchases commodities to operate. Thus, total payments of each activity for commodity inputs are represented by the column sums of the IO table. The payment entries for intermediate consumption are measured at market prices. Since there is no available IO for 1995 that spans the array of activities of interest in this study, it is necessary to construct a new IO on the basis of available information. As previously mentioned, the IO column coefficients for the sectors available in the 1995 published IO table are made as consistent as possible with the data published in the national accounts and with the Embrapa technological coefficients where available. A judgment has to be made for the intermediate demand for newly introduced agricultural activities: it is assumed that the coefficients are equal to those of reasonably similar technologies; otherwise the coefficient for other agriculture is used as an input times the share of the new agricultural activity relative to the other agriculture activity in the IBGE IO table.

For the agricultural activities that are not available in the IBGE IO table and for which no technological coefficients are available, it is assumed that the column coefficients are the same as for the activity called other agriculture in the IBGE IO table.

In the context of the SAM, trade and transportation activities provide inputs (in the row) to the commodities column. Thus, goods at the farm and factory gate are transformed into goods that form part of total supply by including marketing and transportation to the stylized national commodity market.

The value added matrix in the SAM shows labor, land, and capital in the rows, and activities in the columns. Each activity purchases labor and capital to operate alongside intermediate inputs. The value added entries are measured at factor cost. Agricultural activities employ agricultural labor from their region, while nonagricultural activities, including trade and food processing, employ nonagricultural labor. The labor and capital use data have to be disaggregated for the agriculture sectors along the activity row, as is the case for intermediate consumption inputs. For other activities, data are immediately available, due to the low level of disaggregation. Total labor, land, and capital value added are allocated across the agricultural activities based upon the agricultural census.

The output taxes vector shows output taxes paid by each activity. In the 1995 national accounts, data on output taxes only exist for 25 nationwide activities. Further disaggregation of the agricultural activities is therefore needed to go to the regional level and for the agricultural activities that are previously not included. For the agricultural sector as a whole, output taxes were negative indicating subsidies to the sector. These subsidies are relatively small, amounting to about 1 percent of the value of agricultural production. The subsidies are allocated across regional agricultural activities according to activity shares in sector sales.

The domestic sales cells are also referred to as the "make matrix," as it is here that the results of individual activities (in the rows) are combined to form domestic supply of marketed commodities (in the columns). Domestic sales are identical to total sales values (the row totals). Since there are four regional agricultural activities mapping into each agricultural commodity, domestic sales for these commodities are calculated by summing the row totals for the corresponding agricultural activities.

The domestic sales also contain information on marketing margins. These margins are from transport costs as well as wholesale and retail trade margins. For local production destined for the domestic market, they represent the difference between the factory or farm gate and consumer prices. Margins enter each column of the domestic sales matrix along the trade and transportation activities rows. National accounts data provide information on marketing margins, but they do not discriminate between margins associated with activities being located in different areas of the country. Since the four regions taken into consideration in the analysis have varying degrees of infrastructure, regional marketing margins are important. These regional margins are estimated by calculating the average distance to the closest market and using the ratio of these values relative to the industrial South to multiply the trade and transportation coefficients of each agricultural sector as obtained from transportation cost surveys (ESALQ 1998).

Imports and import tariffs appear, respectively, in the government row and in the rest of the world row under the commodity columns; there are 2 rows and 44 columns with entries pertaining to imports and import tariffs. The national accounts data give sufficient information to establish the imports and tariffs for all of the commodities of interest except the nine agricultural commodities that are separated out of other agriculture. Data for these commodities are obtained from the INTAL database available on-line through the Inter-American Development Bank. A similar approach is taken with the export of these commodities.

The SAM contains several household categories to support the analysis of the welfare implications for different income groups of different policy scenarios. The categories are high income, urban medium income, urban low income, rural medium income, and rural low income. The information on household characteristics is incorporated by using the results of the 1996 National Household Survey (IBGE 1997c; IBGE 1997d). The private consumption of the marketed commodities matrix shows commodities in rows and consumption values for each household column. In the same way, the government consumption vector shows commodities in rows down the government column. Currently, these entries reflect total consumption of services by government. The savings and investment demand vector appears in the commodities rows down the capital column. In the national accounts and in BRASAM investment and changes in inventories are treated separately.

The labor column in MACSAM is disaggregated into agricultural and nonagricultural labor. It is further differentiated as skilled or unskilled labor, while the capital column is divided into several land categories and capital; agricultural labor and land categories are specified regionally. Gross profits for the agricultural activities are allocated to the capital row, except for a share going to land based on the return to land being used by the activity (FGV1998).

The remaining entries in the raw BRASAM correspond exactly to the entries in MACSAM. These are scalar entries that require no disaggregation.

To achieve a strict balance in BRASAM, which is required by construction, a minimum cross entropy-balancing procedure is applied, as set forth in Golan, Judge, and Robinson (1994). In the next section a summary of the structure of the Brazilian economy is presented with data from the balanced version of BRASAM.

An Overview of the Brazilian Economy: Regional Production and Income Distribution

Even though the objective of this research is to represent the interactions in the Brazilian economy as a whole by focusing on deforestation and agricultural economic development, the analysis is inextricably linked to the agricultural sectors of the Brazilian economy. Table 4.4 shows that agriculture contributes 10 percent to net national income (value added)—14 percent if the food processing sectors (4 percent) are included. This relatively small share is consistent with the common view that Brazil is an economy with a well-developed industrial sector (21 percent), and services sector (44 percent includes government services). For the purposes of this report, it is interesting to observe that the marketing margins expressed by trade and transport add up to 11 percent of total value added. This implies that approximately one-tenth of the market price paid for the average commodity can be ascribed to transportation and transaction costs. This number can be misleading because it is a national average, whereas transportation costs and trade margins are strongly dependent on location. This is particularly relevant to this research project because at transportation costs are much higher in the Amazon region than in the South/Southeast region, where infrastructure is very well developed.

Agriculture (along with services) stands out as the sector with the highest ratio of value added to output (approximately equal to 0.6). This means that it is the sector in which households (as the owners of the factors of production) get the highest return for each dollar of output produced. Agriculture appears then to be a good income-generating sector. One should, however, take care in drawing conclusions from this ratio: first, some sectors are subsidized, thereby inflating the value added measure by distorting the price system; second, Brazil is known to have one of the worst income distribution statistics in the world, and this is particularly true in rural areas where land ownership is concentrated among a few large landowners. For these reasons, the initial statement has to be qualified in order to examine poverty alleviation options. An in-depth analysis of agricultural activities is required—one that takes into consideration that different activities in different regions have different effects on income distribution. For example, Gasques and Conceição (1999) report that land ownership is less concentrated in the Amazon than in the Northeast and Center-West regions of Brazil.

Table 4.5 provides a schematic representation of the structure of the Brazilian commodity markets at a national level. It shows that more than one-third of the total value of agricultural production occurs in the meat and dairy sectors (R$35 billion). Production of annuals has the same order of magnitude (R$36 billion). The remaining production is distributed among perennials, logging, and other agriculture (mainly fishing), with coffee standing out as the major perennial crop.

The processed food sector is very important in the context of this analysis because it processes the major agricultural export products such as coffee and sugar.

Table 4.4 Value-added structure for Brazil, 1995

Sector/region	Value (R$ billion)	Share of GDP (%)
Agriculture		
Amazon	4.97	0.01
Northeast	8.96	0.02
Center-West	6.14	0.01
South/Southeast	33.37	0.06
Total agriculture	53.44	0.10
Nonagricultural sectors		
Processed food	21.73	0.04
Oil and mining	11.44	0.02
Manufacturing	116.15	0.21
Construction	50.30	0.09
Trade and transportation	61.24	0.11
Services	247.49	0.44
Total nonagriculture	508.35	0.90
Total value added	561.79	1.00

Sources: IBGE 1997a, 1998.

These products are exported only after processing. For this reason, many agricultural products in Table 4.5 do not appear to have exports; in fact, they are intermediate inputs to an export-producing sector (R$11 billion from the export of processed foods). Soy is the major crop to be exported in part in its unprocessed form, along with products from the other perennials and other annuals categories. On the import side, small import shares relative to output are the rule, with the exceptions of other annuals, which includes wheat, and forest extraction, which has considerable two-way trade because of the heterogeneity of the goods included under it. Analogous to agriculture, the food-processing sector can be identified as an export-driven sector with small import shares. This is not true of the other nonagricultural sectors, which either exhibit strong two-way trade (oil and mining and manufacturing, for example) or are nontraded goods (construction and trade and transportation).

As a first step toward understanding the regional structure of Brazilian agriculture, Table 4.6 subdivides farms into large and small operations according to whether they exceed 100 hectares (with the exception of the Amazon where a large farm is assumed

Table 4.5 Structure of the Brazilian national commodity markets, 1995

	Sectoral values (R$ billion)				Ratios (%)	
Sector/commodity	**Output**	**Exports**	**Imports**	**Sales**	**Export/output**	**Import/sales**
Agriculture						
Coffee	5.03	...	...	5.03	0.00	0.00
Cocoa	0.53	0.06	0.01	0.48	0.12	0.02
Maize	6.24	0.01	0.13	6.36	0.00	0.02
Rice	3.11	...	0.04	3.15	0.00	0.01
Beans	2.09	...	0.10	2.19	0.00	0.04
Manioc	2.53	...	...	2.53	0.00	0.00
Other perennials	6.57	0.44	0.30	6.43	0.07	0.05
Other annuals	7.65	0.87	1.01	7.79	0.11	0.13
Sugar	8.49	...	...	8.49	0.00	0.00
Soy	3.83	0.73	0.06	3.16	0.19	0.02
Horticulture	1.98	0.06	0.02	1.94	0.03	0.01
Milk	10.47	...	...	10.47	0.00	0.00
Cattle and swine	17.58	...	0.20	17.78	0.00	0.01
Poultry	7.03	0.07	0.03	6.99	0.01	0.00
Forest extraction	0.41	0.10	0.25	0.56	0.23	0.42
Logging	4.21	0.15	0.07	4.13	0.04	0.02
Deforestation	0.60	...	...	0.60	0.00	0.00
Other agriculture	4.47	0.01	0.83	5.29	0.00	0.16
Agriculture subtotal	92.82	2.50	3.05	93.37	...	...
Nonagricultural sectors						
Processed food	147.49	11.37	2.55	138.67	0.08	0.02
Oil and mining	35.75	3.19	3.86	36.42	0.09	0.10
Manufacturing	373.13	23.32	37.55	387.36	0.06	0.10
Construction	102.80	...	...	102.80	0.00	0.00
Trade and transportation	118.69	3.59	2.60	117.70	0.03	0.02
Services	397.89	2.34	5.45	401.00	0.01	0.01
Nonagriculture subtotal	1,175.75	43.81	52.01	1,183.95	...	...
Total	**1,268.57**	**46.31**	**55.06**	**1,277.32**	**...**	**...**

Sources: IBGE 1997a, 1998a.
Note: The leaders (...) indicate a nil or negligible amount or "not applicable."

to exceed 200 hectares.). Animal production plays a prominent role throughout Brazil, cutting across farm sizes. Value added in animal production, depending on the region, accounts for 42 to 62 percent of small-farm value added. For large farms, this ranges from 40 percent for South/Southeast to 86 percent in the Amazon. Other activities, such as annuals and perennials, vary in importance depending on the region and farm size. In the Amazon, production of annuals is important to smallholders but not to large farm enterprises. Quite the opposite is true in Center-West, where annuals are important to large farms (due to soy production) but not to small farms. Production of perennials is economically relevant to both farm sizes in the Northeast and South/Southeast and to small farms in the Amazon, whereas it is virtually absent in the Center-West.

It is worthwhile to point out the commodities being produced by the economically relevant activities (see Appendix B, Table B.1). Production of annuals by smallholders in the Amazon is geared mainly toward manioc, rice, and beans. In the Northeast, smallholders produce those same staple goods, but maize, other annuals, and horticulture also constitute a considerable share of annuals production. The more diversified production in the Northeast relative to the Amazon may be explained by the different soil characteristics or by risk-spreading behavior adopted by small farms in the drought-prone Northeast. The main annual crop for large farms in the Northeast is sugarcane. In the South/Southeast, production of annuals is quite diversified, both at the small- and large-farm levels, with the small farms mainly producing maize, horticultural goods, and other annuals, and large farms producing sugarcane, maize, rice, soy, and other annuals.

With respect to perennials, smallholder production in the Amazon and the Northeast largely consists of other perennials (which include mango, avocado, papaya, coconuts, bananas, citrus, apples, pears, and the Amazonian fruits, *cupuaçu* and *caju*). Coffee, traditionally an important sector in the South/Southeast (for both small and large farms), has become an important product in the Amazon region with the development of coffee-producing areas in Rondônia. Coffee contributes as much as 23 percent of the value of smallholder production of perennials in the region. The remaining tree crop, cocoa, is produced mainly in the Northeast by both small and large farms.

Animal product activities on large farms generally focus on beef and pork production, followed by milk production and poultry. Milk's share is larger at the small-farm level than the large. There is also more regional variation in terms of what is produced: for example, poultry constitutes an important share of animal products in the Northeast and South/Southeast but not in the other two regions.

According to data from IBGE (1998a), approximately 5 million farm enterprises exist in Brazil. Of these, 47.9 percent are in the Northeast, 38.1 percent in the South/Southeast, 9.3 percent in the North[30], and 4.9 percent in the Center-West. Most of these farms (74 percent) are operated by the

Table 4.6 Value added of regional agricultural output, differentiated by producer size (R$ billion)

Commodity	Amazon	Northeast	Center-West	South/Southeast	National
Small farm products					
Annuals	0.82	1.57	0.17	6.53	9.09
Perennials	0.37	0.86	0.04	3.40	4.67
Animal production	1.32	2.33	0.73	8.64	13.02
Other agriculture	0.23	0.77	0.23	0.32	1.55
Subtotal	2.74	5.53	1.17	18.89	28.33
Large farm products					
Annuals	0.12	1.02	1.22	5.54	7.90
Perennials	0.06	0.47	0.05	2.40	2.98
Animal production	1.32	1.38	3.13	4.99	10.82
Other agriculture	0.05	0.21	0.45	0.07	0.78
Subtotal	1.55	3.08	4.85	13.00	22.48
Forest products	1.33	0.34	0.12	1.47	2.60
Total	**5.62**	**8.95**	**6.14**	**33.36**	**53.41**

Sources: Author's estimation of Social Accounting Matrix based on IBGE 1997a, 1998a.

[30]Some data are more readily available for the North than for the definition of the Amazon in this report. For this reason, and given the large overlap between the two geographic specifications, the North is used here as a proxy for the Amazon when farm and income distribution numbers are presented in the context of a qualitative discussion.

Table 4.7 Farm establishments by size and land ownership concentration

Region	Total number of farm establishments	Share of total regional establishments by farm size (%)		
		Small	Medium	Large
Amazon	443,570	0.91	0.07	0.02
Northeast	2,309,085	0.94	0.05	0.00
Center-West	242,220	0.59	0.32	0.08
South/Southeast	1,843,308	0.89	0.10	0.01
Brazil	4,838,183	0.90	0.09	0.01
		Share of total regional land area by farm size (%)		
	Land area (1,000 hectares)	Small	Medium	Large
Amazon	58,359	0.26	0.22	0.52
Northeast	78,296	0.30	0.40	0.30
Center-West	77,567	0.05	0.25	0.70
South/Southeast	108,446	0.31	0.44	0.25
Brazil	322,668	0.24	0.34	0.42

Sources: Agricultural Census, 1995/96 (IBGE 1998a).

Note: Farms are categorized here as small (less than 100 hectares), medium (100 hectares to less than 1,000 hectares), and large (more than 1,000 hectares).

owners, while renters and sharecroppers operate 11 percent of farms and squatters the remaining 15 percent. The regional distribution of land reported in the lower half of Table 4.7 does not have a direct relationship with the generation of agricultural value added reported in Table 4.6. Such regional differences are not surprising given land values and agricultural factor use (land, labor, and capital), which are very region-specific (Appendix B, Table B.2).

While in all regions a broad majority of producers operate small farms, medium and large farms account for most of the land, indicating an unequal land distribution throughout Brazil. Estimates by Gasques and Conceição (1999) of the Gini Index for land concentration in the different regions of Brazil indicate that land ownership has historically been concentrated among a small group of wealthy landowners (Table 4.8). The Northeast remains the region with the most unbalanced land distribution, while at the other extreme the South and Southeast regions have relatively lower Gini coefficients. The Northern region, which includes a large part of the Legal Amazon, has apparently entered a period of reconcentration in land ownership, after having had a steady decrease in the Gini coefficient during 1975–85.

The government of Brazil has tried to address the inequality in land ownership by introducing gradual land reform; by September 1998, it had allocated land to 359,000 families on an area of more than 17 million hectares. Even so, the potential demand for land reform exceeds by far what the government has been able to provide: depending on the methods used, estimates for the number of "clients" of land reform usually vary from 2.3 million to 4.5 million families, requiring up to 160 million hectares.

Land reform is relevant to the Amazon and to deforestation because it has relied heavily in the past on land in the Legal Amazon. This can be seen in Table 4.9, which illustrates the number of families that were settled by the Instituto Nacional de

Table 4.8 Evolution of land ownership concentration: The Gini coefficient, 1970-1995

Region	1970	1975	1980	1985	1995
North	0.831	0.863	0.841	0.812	0.820
Northeast	0.854	0.862	0.861	0.869	0.859
Center-West	0.876	0.876	0.861	0.857	0.831
Southeast	0.760	0.761	0.769	0.772	0.767
South	0.725	0.733	0.743	0.747	0.742

Source: Gasques and Conceição, 1999.

Colonização e Reforma Agraria (INCRA): before 1985 approximately 70 percent of the families being assigned land were located in the Northern region, highlighting how initially the land reform program was oriented toward opening new lands rather than redistribution. The role of the Amazon agricultural frontier has subsequently declined to 36 percent of families during the period 1986–94, and then to 22 percent during 1995–96 period. This coincided with a greater emphasis on land reform in the Northeast, implying a shift toward reallocation of existing agricultural land rather than relying on frontier areas. Nonetheless, the role the Amazon has played in providing opportunities to the landless (by providing cheap land) must not be overlooked. This highlights the potential conflict that may arise in the future between income distribution, environmental objectives, and the government's budgetary constraints if the current estimates requiring160 million hectares to complete the land reform process are correct.

Migration Patterns Inside Brazil

Background

Land reform can be viewed as only one among a number of factors influencing internal migration in Brazil, whether rural–urban or rural–rural migration. Therefore, it is necessary to understand the broader context in which migration occurs.

According to Perz (2000), the rural exodus in Brazil revolves around the capitalization of agriculture, employment growth in urban industries, and the high rate of natural increase in rural populations (excluding migration). These factors, however, have had different roles in different periods of recent Brazilian economic history. Between 1970 and 1980, there was a state-led

Table 4.9 Number of families that have benefited from land reform projects

Region	Up to 1985	1986-94	1995-96	Total
North	9,287	29,636	12,300	51,223
Northeast	1,835	31,571	28,878	62,284
Center-West	897	5,369	1,845	8,111
Southeast	402	5,748	2,191	8,341
South	825	10,446	10,478	21,749
Brazil	13,246	82,770	55,692	151,708

Source: Convênio INCRA/CRUB/UNB 1997.

economic expansion with all three factors contributing to rural–urban migration; during that period 38 percent of the 1970 rural population moved to urban areas (Goldin and Rezende 1990). Between 1980 and 1985, when a debt crisis led to a withdrawal of state incentives to large farms, an industrial downturn, and an urban wage decline, it seems likely that rural-urban migration slowed (Perz 2000).[31] Macroeconomic reforms, following in the wake of the debt crisis, had a positive influence on agro-industrial expansion: (1) an exchange rate devaluation of the Brazilian currency made agricultural exports more profitable, (2) trade policy favored processed agricultural goods through tax exemptions, and (3) the government introduced minimum price guarantees to reduce the uncertainty of returns to agriculture (Goldin and Rezende 1990).

Demographic data for 1991 and 1996 indicate that the rural population declined from 36 million to 34 million during that period. While the debt crisis may have temporarily slowed rural-urban migration, it seems that the recovery of the urban economy and the increasing importance of agro-industrial exports after the crisis caused the flow of migrants from the countryside to continue.

Migrants from rural areas might go to urban areas or to the agricultural frontier. As Martine (1990) points out, frontier growth has become progressively less meaningful in terms of population absorption. But even though only a small share of total migrants chose the agricultural frontier, in labor-scarce areas of the Amazon, they could have a considerable impact on deforestation rates by providing the labor necessary to cut down trees. Table 4.10 shows the gross migration between regions from 1991 to 1996.

To understand the migration patterns within Brazil, it is important to recognize substantial regional differences in agricultural technologies adopted, infrastructure, and relative sizes of rural population. In the Legal Amazon, as more and more land was used to raise cattle, diminishing labor requirements along with rapid urban growth spurred a substantial intraregional rural-urban urban shift. However, the Legal Amazon also attracted many rural-rural migrants from other regions, which somewhat counterbalanced this effect.

Many came from the Northeast, where the rural population is large and income distribution historically skewed. More than 40 percent of all net rural outmigration to

Table 4.10 Gross migration between regions inside Brazil, 1991–1996 (number of persons)

Region	Amazon	Northeast	South/Southeast	Center-West	Urban Brazil	Total outmigration
Amazon	242,840	5,193	9,276	4,778	194,368	213,614
Northeast	71,171	442,074	53,223	18,679	901,795	1,044,868
South/Southeast	34,603	53,128	753,738	23,820	1,400,710	1,512,261
Center-West	28,496	6,773	18,788	133,703	256,472	310,529
Total inmigration	134,270	65,093	81,287	47,277	2,753,345	

Source: Author's estimates based on IBGE 1998b.

Notes: Internal migration to urban areas in Brazil is based on the assumption that 32 percent of urban inflow is associated with rural outmigration. It is meant as an approximate number, and in this sense it is consistent with data from Perz (2000) 1986-91and with early work by Reis and Schwartzman (1978).

[31]Limited data availability for the early 1980s makes it difficult to come to a definitive conclusion on this issue.

other regions of Brazil can be attributed to the Northeast. The Southeast, with a large urban population and an expanding services sector, has consistently been the main rural–urban receiving area, while the South has sustained a mainly localized rural–urban shift (Perz 2000).

To consider migration mechanisms in this analysis, one must identify the determinants of internal migration in Brazil. At the aggregate level, several studies were carried out during the 1960s and 1970s that attempted to relate regional and sectoral wage differentials and internal migration. Sahota (1968), using an econometric model, measured the responsiveness of migration to differentials in earnings and other variables. In the same vein, Graham and Buarque de Holanda (1971) estimated net migration for each state from 1872 to 1970 and found a significant association between relative state income and rates of migration. A number of studies analyzing migration at a more local level have shown that a broad mixture of "push" and "pull" factors is necessary to explain the decision to migrate (Brito and Merrick 1974; Duarte 1979; Perlman 1977). However, as Martine (1990) points out, both aggregate and local survey data show the predominance of economic motives of migration in Brazil.

Estimation of Migration Thresholds

After constructing the model and performing an initial set of sensitivity analysis runs, it became apparent that interregional migration paths in the model have a major impact on a subset of the simulations (in particular the ones linked to macroeconomic or interregional changes).[32] There is almost no recent literature on internal migration in Brazil, the exception being a survey by SENAR/FGV (1998) investigating the propensity to migrate to urban areas. The study examines the propensity to migrate between families and their offspring in different regions (South, Southeast, Center-West, and the states of Pernambuco and Ceará), but no attention is paid to rural-rural migration or to the economic determinants of migration. Since the research presented here focuses on the economic determinants of migration to the agricultural frontier, the estimation must first determine rural areas of origin and destination and then link migration to differences in income. For this reason, wage differential threshold parameters are introduced to characterize migration mechanisms, and they are estimated using data on interregional wage differentials and migration from one rural area to another and between rural and urban areas.

The wage differential threshold parameter indicates how much the relative interregional wage differential for a factor must shift between two regions before migration from one region to the other begins to occur. The principle behind this approach is that migration to certain regions may be preferred over others. It is hoped that estimation of these thresholds will capture a diverse set of motivations that may affect the decision to migrate, such as the risk involved in moving to an area (Harris and Todaro 1970), family support networks in the receiving region, or simply climate and infrastructure conditions of the receiving region relative to the area of origin.

For the purpose of the threshold parameters estimation, it is assumed that migration between two regions is described by a piecewise-linear relationship between the interregional wage differential and the number of people migrating (Figure 4.1).

[32]In the initial version, the model either allowed migration with full wage equalization or migration was precluded a priori. This led to migration occurring even for simulations with relatively small changes in interregional wage differentials.

Figure 4.1 Migration as a function of wage differentials between regions

Migration from region A to B

(Wage differential between A and B)

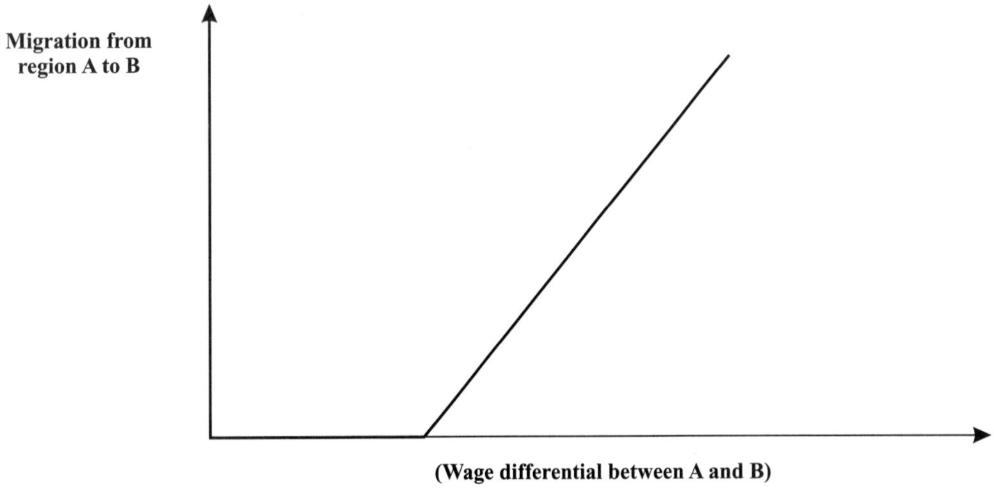

The relationship in Figure 4.1 can be expressed as

$$MIGRSH_{f_1,f_2} = promig_{f1} \cdot (DWG_{f_1,f2} - dwt_{f_1,f_2}),$$

where the share of population migrating from f_1 to f_2 ($migrsh_{f_1f_2}$) is obtained by multiplying the propensity to migrate of the population in f1 ($PROMIG_{f_1}$) by the excess wage differential between the two regions (DWG) relative to the threshold (dwt_{f_1,f_2}). This expresses a disequilibrium adjustment process, which causes factor returns to converge. It is completed when the wage differentials between regions are equal to the thresholds. To be able to use this expression to characterize migration between Brazilian rural regions and rural–urban migration, the thresholds and the propensity to migrate out of regions have to be estimated.[33]

To estimate the parameters, a cross entropy estimation method is adopted as presented in Golan, Judge, and Miller (1996). The problem is to find a new set of *DWT* coefficients that minimize the entropy distance between an assumed prior $\overline{DWT}$ and the new estimated coefficient matrix. The aim is to minimize the expected information value of additional data, given what is known (sample and prior). In mathematical terms the problem can be presented as

$$\min \left[\sum_{f_1,f_2} DWT_{f_1,f_2} \ln \frac{DWT_{f_1,f_2}}{\overline{DWT}_{f_1,f_2}} \right].$$

It is subject to

$$migrsh_{f_1,f_2} = PROMIG_{f1} \cdot (dwg_{f_1,f2} - DWT_{f_1,f_2})$$

$$\sum_{f_1,f_2} DWT_{f_1,f2} = 1 \text{ and } 0 \le DWT_{f_1,f2} \le 1,$$

with $PROMIG\,f_1$ and $DWT\,f_1f_2$ being the parameters to be estimated. The estimation data focus on migration occurring between

[33]The model used for the analysis in this report analyzes the movement from one equilibrium to another following a shock or a structural change in the economy; therefore, the wage differential thresholds (rather than the propensity to migrate once the threshold is exceeded) will be central in determining the outcome of the scenarios being analyzed.

1991 and 1996. Table 4.11 includes the population shares that migrated either between rural areas or from a rural area to an urban setting and the average wage differential during the period.

To compute the share of the population migrating, the net migration flows are obtained from Table 4.10. The results, which are presented in Figure 4.2, highlight the population size in each category (size of circle) and the volume of migration between categories (along the arches).

The solution to this problem would typically be obtained analytically by setting it up as an unconstrained optimization problem; however, the problem has to be solved numerically because no closed-form solution exists. The outcome combines the information from the data and the prior DWT. For the prior, a uniform wage differential of 10 percent is assumed before migration occurs. If the data are noninformative, then the solution will simply coincide with the prior. In this case, however, the results from the estimation deviate substantially from the prior. This indicates that a wage differential threshold for migration from any agricultural region to an urban environment between 7 and 8 percent is substantially lower than the prior (10 percent) (Table 4.12).

What also emerges is that migrants from the Northeast make strong distinctions among regions, as expressed by the smaller

Table 4.11 Average wages and population for each region

Region	Average monthly wage (R$)	Population (millions of people)
Rural Amazon	216	4,249
Rural Northeast	169	15,575
Rural Center-West	200	1,636
Rural South/Southeast	185	12,534
Urban Brazil	280	123,078

Sources: Brazil, Ministry of Labor 2000; IBGE 1998b.

Figure 4.2 Net migration flows between regions used in the estimation of migration functions.

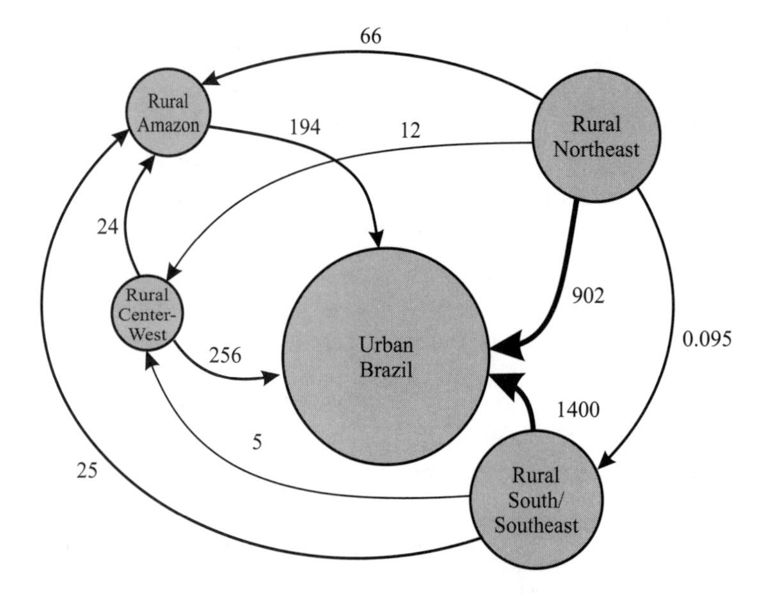

Notes: Units are thousands of people migrating (size of circles indicates approximate size of population at a node, the thickness of lines indicates magnitude of migration flow between two nodes).
Net migration flaws were computed based on Table 4.10

Table 4.12 Propensity of unskilled labor to migrate and wage differential threshold (%) before movement occurs between two regions

		Migrant destination wage differential threahold (%)				
Migrant origin	Propensity to migrate from origin	Urban	Rural Amazon	Rural Northeast	Rural Center-West	Rural South/ Southern
Rural Amazon	0.30	7.70	...	...	...	...
Rural Northeast	0.18	7.10	19.40	...	15.10	8.60
Rural Center-West	0.76	7.90	5.50	...	...	...
Rural South/ Southeast	0.42	7.60	13.90	...	7.40	...

Notes: The leaders (...) indicate a non-applicable entry because no net migration was observed from origin to destination. Estimation results are based on data in Tables 4.10 and 4.11, using the method described in the text.

wage increase required as an incentive to move to preferred areas. They prefer to migrate to urban areas (7.1 percent wage differential), followed by migrating to agricultural areas in the South/Southeast (8.6 percent), and the Center-West (15.1 percent). The Amazon is a distant last choice as a destination for Northeastern migrants, requiring a 19.4 percent wage differential before migration along this route begins to occur. In other words, migrants would only consider migrating to the Amazon if they thought they could increase their wages by 19.4 percent. Interestingly, the Amazon is the preferred destination for migrants from the sparsely populated Center-West, where migrants require a wage differential threshold of only 5.5 percent. This may be a reflection of the fact that Center-West is attracting migrants from the South/Southeast (who require only a 7.4 percent differential) and the Northeast (15.1 percent) and in the process its agricultural frontier is being pushed into the Amazon.

Further information that is obtained from the estimation (which is not used in the CGE model) is the propensity to migrate when out of equilibrium. This highlights the fact that regional differences are not fully accounted for by the wage differential thresholds. Instead, once these thresholds are exceeded, the extent of migration occurring relative to the excess wage differential will differ by region of origin. Similar to the SENAR/FGV (1998) study, this study finds that the Northeast has the lowest propensity to migrate, with Center-West and South/Southeast having considerably higher propensities and the Amazon falling somewhere in between (first column in Table 4.12).

Agriculture and Land Degradation Processes in the Brazilian Amazon

Background

Biophysical processes related to crop sustainability are important topics for farmers in the Brazilian Amazon. Among these processes are some that can substantially reduce agricultural productivity. When a plot is cleared, soil degradation and weed infestation begin to appear after just a few cropping cycles. Land degradation affects the stocks of available agricultural land, thereby affecting agricultural producers' decisions, especially given that different productive activities will require different land types and have different impacts over time on land quality.

The framework presented in the modeling section is a first step in linking biophysical changes occurring with present land uses to the economic incentive for agents to modify existing land use patterns. Including a representation of physical processes in the economic framework is important because these processes are a major constraining factor for regional development in the Amazon region.

Continuous time and discrete time models of these processes can be used interchangeably with similar results (Shugart, Crow, and Hett 1973). In most cases, empirically based models use estimates of change determined by resampling the landscape at discrete time intervals. The model is in discrete time, and the intervals considered are years. This is the approach adopted for the estimation here. The state space is also discrete because a finite number of states in which land can be classified are considered.

Estimation of the Land Transformation Matrix

For a first-order stationary Markov process with exogenous land use, the transition probabilities are usually derived from a sample of transitions (conditional on land use) occurring between two points in time. Depending on data availability, the transition probabilities can be estimated using different data sources. The results here rely on data collected through farm surveys by IFPRI researchers in Acre and Rondônia (Vosti, Witcover, and Carpentier 2002). Depending on data availability, the transition probabilities could be estimated using land use maps or by running Monte Carlo simulations using crop models adapted to tropical areas.

The estimation problem can be represented as a network in which the nodes represent the stocks of the different land types and the parameters to be estimated are the flows linking the stocks over time (Figure 4.3). These flows can be interpreted as probabilities once the ratio of flow to stock is obtained. For a regional application, ideally the data should include plot history for a wide variation of farms. In general, however, it is difficult to find longitudinal time-ordered data that describe individual plot movements from state to state. Instead, for each t there might be a limited number of transitions for aggregate land use data (from land use maps) that show either the number of outcomes or the corresponding proportions in each of the Markov states in each time period. Alternatively, farm surveys may have a cross-section of plot level data but as a subjective probability of transition elicited from the farmer rather than time-ordered observations.

An estimation problem is said to be ill posed if there is not enough information in the data to permit the recovery of the transition probabilities by traditional estimation methods. The ill-posed aspect may arise because the data are mutually inconsistent or there are not enough data points. If traditional estimation procedures are used in situations with very few data points, the problem is said to be underdetermined, leading to highly unstable estimates and arbitrary parameters. In a recent book, Golan, Judge, and Miller (1996) suggest a variety of estimation techniques using what they describe as "maximum entropy econometrics," which can be applied when dealing with ill-posed problems. These techniques may turn out to be very useful in the estimation of the land transformation matrix. Indeed, there is a section of their book that analyzes an ill-posed stationary Markov inverse problem. That it would allow the adoption of farmers' subjective probability of transformation as a Bayesian prior in the estimation process makes it even more appealing for this research. These data could not be otherwise incorporated into a standard estimation technique.

A maximum entropy formulation of the estimation problem (similar to that adopted in the section on the estimation of migration thresholds) was developed using reported degradation times for the Amazon from farm surveys and agricultural extension studies as priors. Based on this formulation,

Figure 4.3 Transition network for estimation of transformation processes

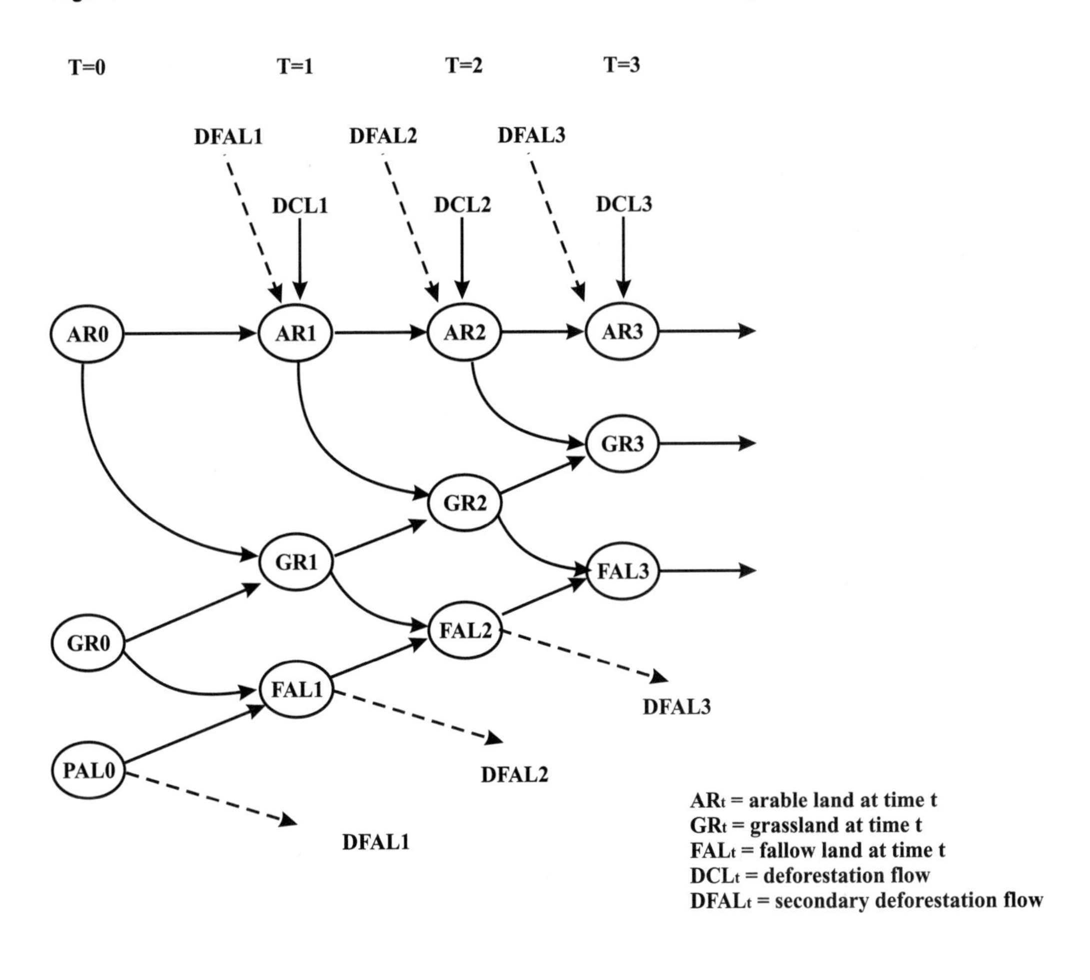

it took 2-4 years for the transition from arable land to grassland under annuals, and 8-15 years to go from grassland under pasture to degraded land that is either abandoned or left fallow (Vosti et al. 2001; Vosti, Witcover, and Carpentier 2002; Fearnside 1997; Weinhold 1999). Selected estimates from these ranges were used as priors and combined with data from the agricultural census for the period 1970–96 at the Amazon-wide level of aggregation, but there was not enough information in the census data (at that level of aggregation) to observe any deviation from the chosen prior. For this reason, an estimate of 3 years was adopted for the transition from arable land to grassland under annuals and 8 years for the transition from pasture to degraded land (which appeared to be reasonable given the literature). Forest recovery (from a farmers perspective) was ignored at this phase of the modeling effort; however, it will be incorporated in the model in the future so as to include fallowed areas in the land use decision process.

Factor Supplies and Elasticities

Deforestation in 1995 was assumed to equal average deforestation between 1992 and 1996 (in hectares). The coefficients for deforestation technology were obtained from Carpentier, Vosti, and Witcover (forthcoming). Timber production in the Amazon and in the rest of Brazil was obtained from the agricultural census. The economic rent to timber was based on a technological specification proposed by Stone (1998). Elasticities of substitution between production factors were taken for industry from Najberg, Rigolon, and Vieira (1995). For agriculture, the substitution elasticity between land and capital was set at 0.4 for smallholders and 0.8 for large farm enterprises. These values are judgment-based estimates, assuming farm enterprises can substitute more easily between factors. The substitution elasticities in the production process of agricultural commodities were obtained through surveys. Sensitivity analysis was performed for the elasticities that were judgment-based, and although in some instances substantial deviations occurred relative to the results reported here, most implications drawn for the different scenarios concerning deforestation and income distribution were found to be valid (see Appendix C for more details).

To conclude this section, several limitations in the data and model formulation must be noted. First, due to the uncertainty surrounding the elasticities, the results of the simulations are meant to clarify the sign and order of magnitude of the effects of regime shifts and should not be interpreted as precise quantitative measures.[34] Second, the model developed here is essentially static, and the results represent the impact of different policy experiments in a timeless world. A dynamic version of the model is being developed, but for the purpose of this analysis, which compares the effects of technological changes in a controlled environment, a comparative statics framework is more appropriate.

[34]To better understand the robustness of the results presented in the simulation sections, a small sensitivity analysis section is included in Appendix C. A full sensitivity analysis, using Monte Carlo simulation techniques will be performed soon and will appear in future publications.

CHAPTER 5

The Effects of Macroeconomic, Interregional, and Intraregional Change

A diverse set of indirect causes–macroeconomic, interregional, and intraregional–has driven deforestation in the Amazon. First, this chapter will consider macroeconomic and interregional aspects of the Brazilian economy that have had immediate policy relevance for deforestation: (1) a devaluation of the real exchange rate, (2) a 20 percent reduction in transportation costs, and (3) technological change in agriculture in the Northeast, Center-West and South/Southeast regions of Brazil.

At the intraregional level, the chapter investigates the impact on income distribution and deforestation rates of (1) different types of technological change in Amazon agriculture, (2) modification of tenure regimes, and (3) fiscal incentives or disincentives introduced to reduce deforestation rates. The main purpose of this comparative exercise is to determine what policy options are available, what trade-offs between environmental and developmental objectives are likely, and whether policymakers should focus on interregional or intraregional policies when considering development and deforestation in the Amazon.

An important distinction when analyzing the underlying causes of deforestation, besides the geographic level of aggregation, is differentiation between types of distortions leading to excessive deforestation. Box 5.1 shows the distinctions between *market failure, policy failure,* and *institutional failure*. It is important to distinguish between these types of failures in order to determine the types of corrective measures required. An overvalued currency and poorly planned infrastructure are examples of potential policy failures, missing markets for environmental services provided by standing forest are an example of market failure, and inadequately specified tenure regimes represent institutional failure.

Crisis and Structural Adjustments: How Macroeconomic Policy Affects Deforestation in the Brazilian Amazon

As of mid-1998 there was speculation, fueled by the recent Asian financial crisis, of a possible devaluation of the Brazilian currency. The agreement reached in October 1998 between the International Monetary Fund (IMF) and the Brazilian government seemed to dispel the uncertainty over Brazil's future. In January 1999, however, the possibility of default by the state of Minas Gerais worsened the already difficult economic position the Cardoso government was facing. The widespread rumor that other states might follow suit sent foreign investors fleeing from the Brazilian capital market. The government, having to choose between making a stand for its overvalued currency or deciding not to intervene, opted in mid-January for a compromise 8 percent devaluation, which was not sufficient to reduce the outflow of capital.

Box 5.1 Market, policy, and institutional failures

Market failure occurs when markets are absent, distorted, or malfunctioning, so that forest goods and services are undervalued or not valued at all. Major sources of market failure include

- externalities in which the effect of an action on another party is not taken into account by the perpetrator;
- missing markets for environmental services and other "open-access" public goods; and
- market imperfections that cause uncertainty, such as a lack of information and knowledge.

Policy failure occurs both when the state fails to take action to correct market failures and when policies are implemented that act as disincentives for sustainable management.

Common examples of policy failures believed by most analysts to encourage deforestation are

- subsidized inputs and credit for land-extensive agriculture and livestock production;
- protection of forest industries through trade restrictions such as log export bans;
- poorly planned transport infrastructure; and
- devaluation, which may encourage agricultural expansion. Extra-sectoral policy impacts, especially those coming from macroeconomic policies or adjustments, give rise to various social, environmental, and economic effects. In many cases these policies may be necessary for a healthy economy. Thus corrective environmental policies are politically complex.

Institutional failure, if institutions are interpreted broadly to include legal rules, organizational forms, norms of behavior, and enforcement mechanisms, can take on many forms, including

- legal specifications that introduce distortions, such as property rights to land being acquired through deforestation;
- weak state control over a territory leading to illegal logging or land clearing; and
- incentives encouraging corruption.

In the end, the government decided to float the exchange rate. The effect was a 70 percent peak nominal devaluation over a period of three weeks. The exchange rate faced an adjustment process that may still be evolving to this day given the uncertain economic situation. It appeared, however, that by mid-2000 the real exchange rate had stabilized at approximately 50 percent of its value relative to 1995.[35] The simulations presented in this section assume that 20–50 percent is a reasonable range for a devaluation (in real terms) once the market adjustment is complete. A series of devaluations that range from 10 to 40 percent are simulated. Results are presented for four different devaluation scenarios, differentiated by the macroeconomic closure describing how

[35]Brazil faces renewed difficulties in 2002, in part because it continues to have a high level of public debt, but also because (1) the 2001–02 crisis in the Argentine economy had a contagious effect in terms of capital flows to the region, and (2) Brazil is in the middle of a presidential campaign whose outcome may produce important economic policy changes.

agents react to the crisis to balance the flows in the economy and by the time horizon assumed. On the macroeconomic closure side, the following definitions apply: balanced adjustment describes a balanced contraction of demand under a financial crisis scenario defined by government consumption and investment spending imposed as fixed shares of total demand. *Capital flight* represents the extreme case in which both the government and consumers do not respond to the crisis, in which case the resulting capital flight acts completely on the investment side of demand. In the first scenario, given the determination of the investment value, the burden of achieving a balance between savings and investment falls on the savings side, affecting the savings rates for the different household categories. In the second scenario, the savings rate and government expenditure are fixed at the initial pre-crisis level, and investment reflects in full the reduction in foreign capital inflows occurring during the crisis.

The scenarios are distinguished by the time horizon of the process of adjusting to

Table 5.1 Factor mobility and utilization for short-run and long-run scenarios

	Factor mobility across activities		Factor mobility across regions or categories		Factor utilization		Adjusting variable	
	SR	LR	SR	LR	SR	LR	SR	LR
Labor								
Rural unskilled labor (all regions)	Y	Y	N	Y	Full	Full	Wage	Wage
Rural skilled labor (all regions)	Y	Y	N	N	Full	Full	Wage	Wage
Urban unskilled								
Food processing	N	Y	N	N	Unem	Full	Utiliz	Wage
Industry	N	Y	N	N	Unem	Full	Utiliz	Wage
Mining	N	Y	N	N	Unem	Full	Utiliz	Wage
Construction	N	Y	N	Y	Unem	Full	Utiliz	Wage
Services	N	Y	N	Y	Unem	Full	Utiliz	Wage
Urban skilled								
Food processing	N	Y	N	N	Unem	Full	Utiliz	Wage
Industry	N	Y	N	N	Unem	Full	Utiliz	Wage
Mining	N	Y	N	N	Unem	Full	Utiliz	Wage
Construction	N	Y	N	N	Unem	Full	Utiliz	Wage
Services	N	Y	N	N	Unem	Full	Utiliz	Wage
Capital								
Rural small farm (all regions)	Y	Y	N	Y	Full	Full	Wage	Wage
Rural large farm (all regions)	Y	Y	N	Y	Full	Full	Wage	Wage
Nonagricultural	N	Y	N	N	Full	Full	Wage	Wage
Land								
Amazon arable	N	Y	Y	Y	Unem	Unem	Wage/ utiliz	Wage/ utiliz
Amazon perennial	N	Y	N	N	Full	Full	Wage	Wage
Amazon grassland	N	N	Y	Y	Full	Full	Wage	Wage
Amazon forest land	N	N	Y	Y	Unem	Unem	Utiliz	Utiliz
Arable land (all other regions)	N	Y	N	N	Full	Full	Wage	Wage
Perennial land (all other regions)	N	Y	N	N	Full	Full	Wage	Wage
Grassland (all other regions)	N	N	N	N	Full	Full	Wage	Wage
Forested land (all other regions)	N	N	N	N	Full	Full	Wage	Wage

Notes: When unemployment (unem) is allowed, factor utilization (utiliz) becomes the adjusting variable. Structural rigidities may also be expressed by a wage threshold: until it is reached, it is the wage that adjusts; beyond the threshold, it is the utilization of the factor that adjusts (these cases are denoted by wage/utiliz). SR is short run and LR is long run; Y is yes, N is no.

Figure 5.1 Logging in the Amazon: Balanced-contraction versus capital-flight scenarios in the short and the long run

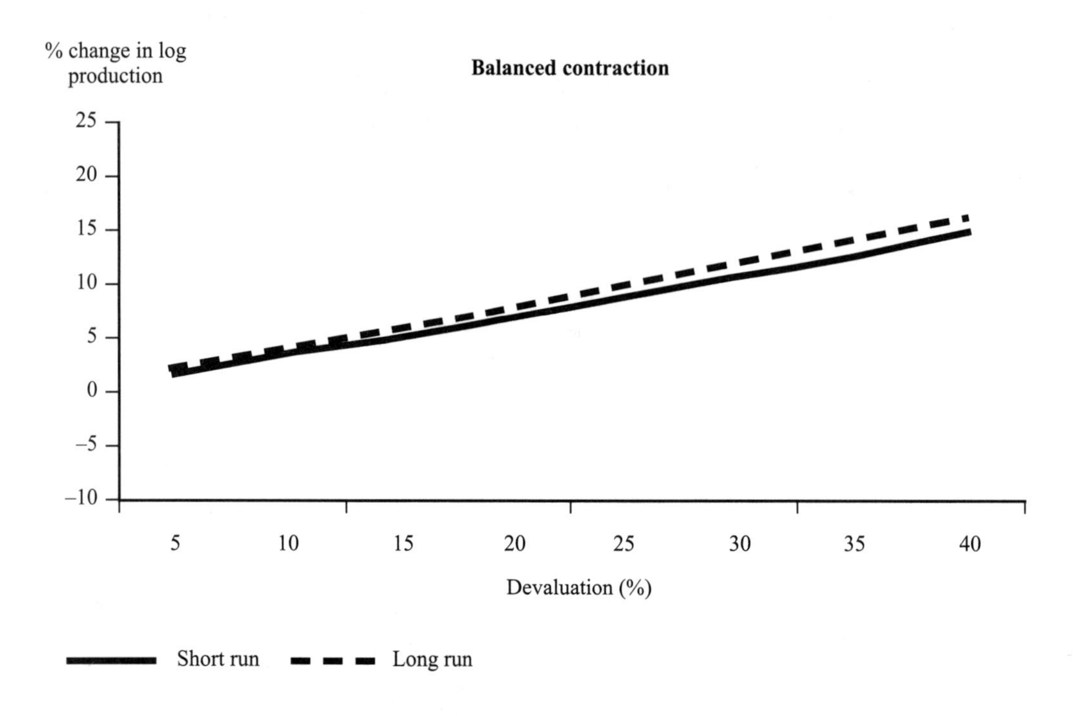

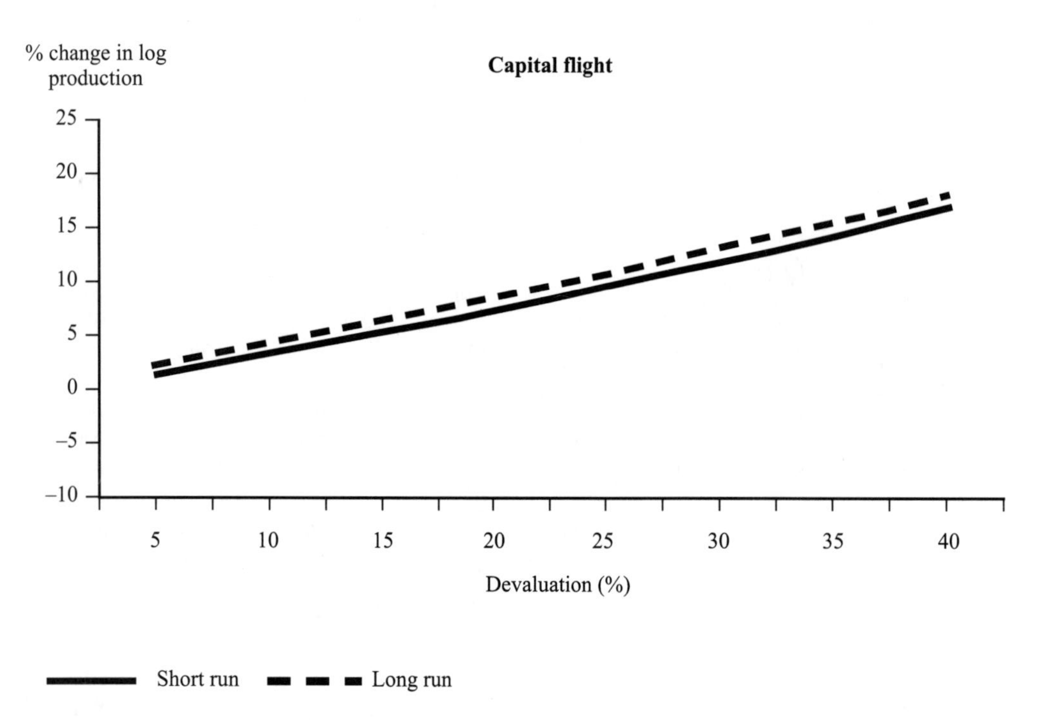

the crisis in the following way: The short-run scenario assumes that wages are rigid, and therefore excess supply in the labor market is possible; it also assumes that in the short run, labor migration between regions is not possible. The long-run scenario assumes wages are flexible and migration between rural areas is allowed. The results presented below are for four scenarios: balanced adjustment in the short run, balanced adjustment in the long run, capital flight in the short run, and capital flight in the long run.

Table 5.1 illustrates the difference in factor mobility and utilization between short- and long-run (SR and LR) scenarios. It is hoped that by considering these four extreme scenarios, any future development arising from the devaluation can be bracketed and a range identified for the values certain critical variables will assume as a result of the devaluation.[36]

Changes in the exchange rate reverberate through the economic system by affecting the relative prices of goods. On the supply side of the economy, prices of export goods rise relative to nontraded goods sold domestically (services and construction, for example). This implies that production shifts toward sectors that produce goods with a high export share. Conversely, on the demand side, the rise in the price of imported goods leads to a greater demand for domestic substitutes of the imported goods. These two countervailing effects lead to a price adjustment on the market for domestically produced goods, which allows all markets to clear. As is to be expected, wages are also affected by this process. The advantage of adopting a general equilibrium framework is that it allows us to take all these processes into consideration at the same time. Given enough microeconomic detail in the model, it is possible to follow the reverberations of a macroeconomic shock throughout the economy—in this case to regional agricultural production sectors and logging. Results of the model's devaluation scenarios are presented in Appendix D, Tables D.1 to D.12.

Both the short-and long-run implications of a devaluation in real terms (as opposed to nominal) of the Brazilian currency on deforestation for agricultural purposes and logging in the Amazon are analyzed (Figures 5.1 and 5.2). In examining these effects, the welfare implications of the devaluation are taken into consideration.

The interesting result that emerges from simulating the devaluation under different macroeconomic closures is how deforestation for agricultural purposes and logging react differently as sectors under the different assumptions on how the economy reacts to the shock. Logging in the Amazon increases uniformly across simulations, with the capital-flight scenario leading to slightly greater increases in logging, compared with the balanced-contraction scenario. This considerable increase in logging arises from a substantial increase in exports from the industry sector that includes processed wood. From a policy standpoint, the only option for avoiding this increase would be to place an export tax on processed wood; however, the price distortion introduced by

[36]The distinction between the short run and the long run is carried throughout the simulation section. As currently modeled, it is only a representation of the two extremes in terms of market rigidities (where one is very rigid and the other is very flexible). In reality, there may be long-term rigidities in factor markets that cause them not to clear even in the long run; however, these rigidities are not easily estimated. The results obtained through a simplified specification of the two extreme cases are meant to encompass other more realistic intermediate situations that may fall between these two extremes.

Figure 5.2 Effects of balanced-contraction versus capital-flight scenarios on deforestation in the short and the long run

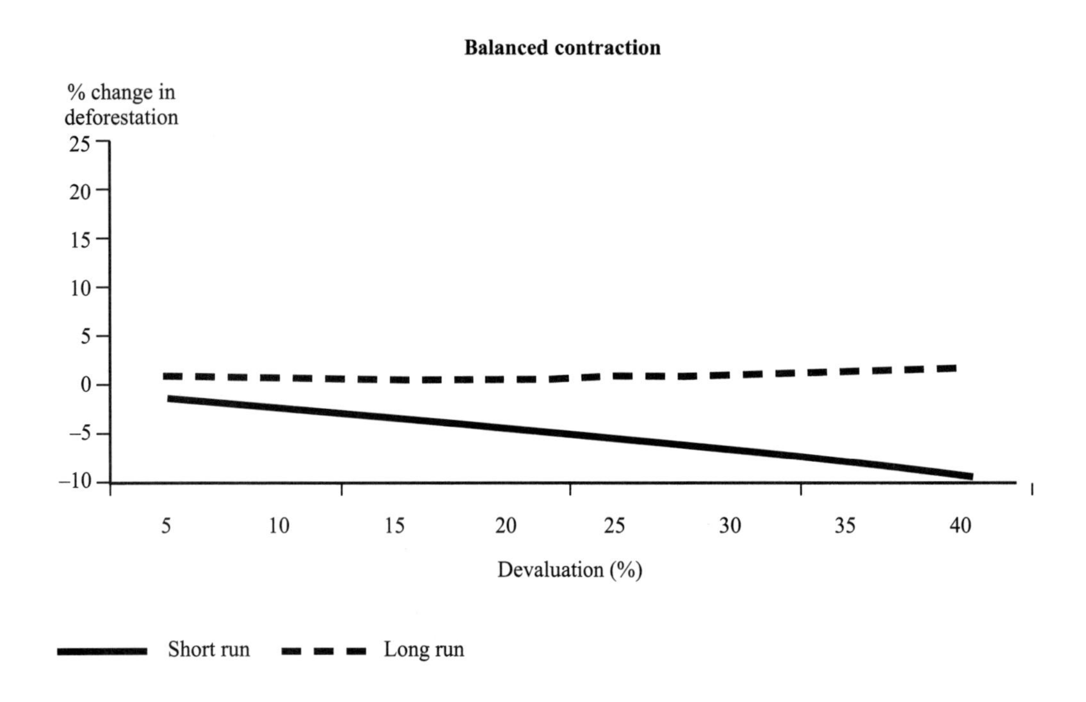

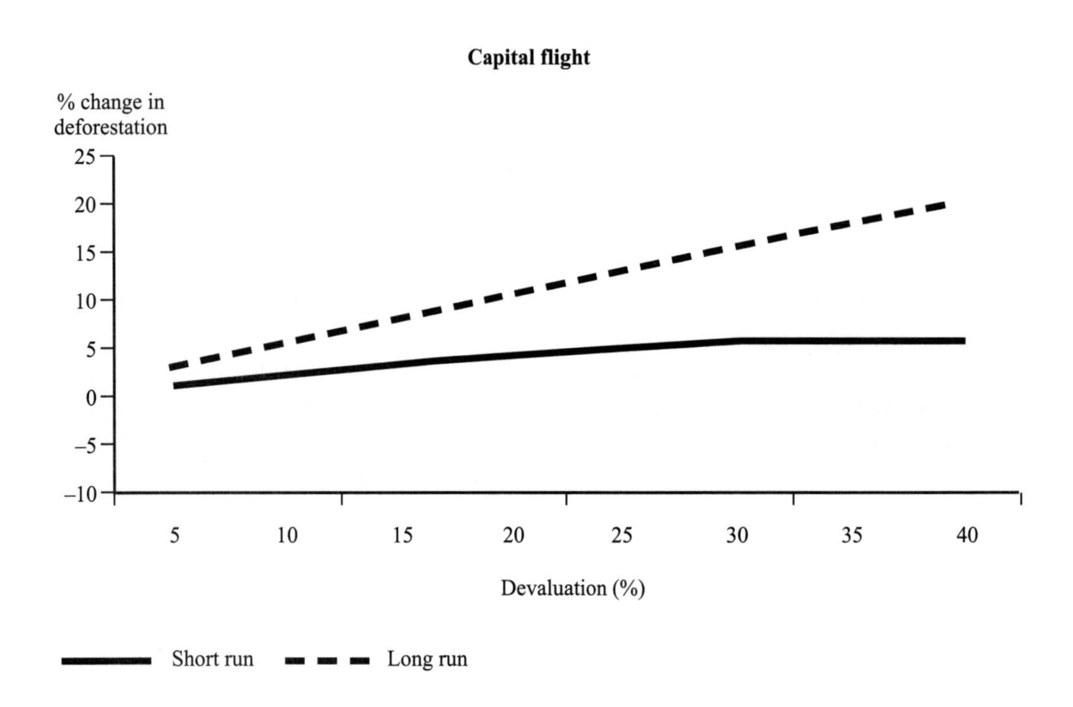

such a tax would have negative welfare effects.

Deforestation to clear agricultural land is quite sensitive to the aggregate behavior of the economy. On the one hand, the balanced-contraction scenario, with a reduction of private consumption, government demand, and investment, would lead to a reduction in deforestation in the short run and a small positive increase in the long run. On the other hand, the capital-flight scenario, where government expenditures and household savings rates are left unchanged (meaning investment will decrease drastically) would lead to a small increase in deforestation in the short run and a substantial increase in the long run. A devaluation of 40 percent causes, in the long run, a 20 percent increase in deforestation, leading to an increase of approximately 4,000 square kilometers in the yearly deforestation rate.

For deforestation, the mechanism underlying the different reactions to the devaluation depends on the returns to agriculture relative to nonagricultural activities. This can best be understood by comparing the changes in output in the long-run scenarios (Appendix D, Tables D.4 and D.10). From the production levels, one can infer that the majority of the agricultural sectors expand more under the capital-flight scenario than under the balanced-contraction scenario. This is a consequence of the strong contraction of the nonagricultural sector as a whole, in particular the sectors producing nontraded goods. The reason this effect is stronger in the capital-flight scenario is that investment is more adversely affected than in the balanced-contraction scenario, thereby causing a drastic contraction of the construction sector. This leads to a larger migrant pool of displaced workers who move into agriculture and in so doing affect the movement of the agricultural frontier in the Amazon. In the short run, without migration, this effect is less pronounced, although the general principle still applies.

The income distribution effects are quite different in the short run than in the long run. In the short run, for both the balanced-contraction and the capital-flight scenarios, rural households stand to gain and urban households to lose from the devaluation. This result is reversed in the long run. The short-run result is intuitive to the extent that agriculture can shift production between exportable and domestically consumed goods, while some nonagricultural sectors cannot (Appendix D, Tables D.2 and D.8). This implies that rural households can adapt better to the crisis than urban households. If agriculture is more attractive, the per capita income of rural households should also increase in the long run (Appendix D, Tables D.5 and D.11). However, this does not occur for a reason that can be explained by taking factor migration into consideration. By allowing labor—but not capital—to migrate from urban to rural areas, it becomes implicit that only the poorest (those without any capital) will migrate to rural areas. This skews the per capita income measure in favor of urban households, because those that stay are the ones that in their income category are not the worst off. The flow of income, in fact, increases for rural areas and decreases for urban areas, but this is not true in per capita terms, since the worst off are the ones who migrate.

Finally, in aggregate growth terms, the crisis affects GDP substantially no matter what course of action is taken. Where the two macro scenarios differ is in who will be affected: if the contraction is balanced, all components of demand will shrink to compensate for the loss in foreign investment arising from the crisis. This balanced macro adjustment limits the long-term effects of the crisis. If, alternatively, the capital-flight scenario takes place, private consumption is not affected much; however, by not adapting the savings rate, future growth will be slower. The dynamic effects of savings and investment decisions are not captured by this model; however, they can be deduced quite easily.

Infrastructure Improvement in the Amazon and the Trade-offs between Environment and Development

In this section the impact of a reduction in transportation costs is analyzed. The policy relevance of such a scenario stems from the changes occurring in infrastructure in the region. The Brazilian government has revisited its policy toward Amazonian development as part of its *Avança Brasil* (Forward Brazil) plan. This is an ambitious program for development in the Brazilian Amazon, involving infrastructure investments of US$45 billion over eight years (1999–2006). The plan emphasizes road paving, river channeling, port improvements, and expansion of energy production. If implemented, it will add 6,245 kilometers of paved highways to the region's road network, including the Santarém–Cuiabá and Humaitá–Manaus highways, which cut through the core of the region. Combined, the two roads cut through 1,800 kilometers of forests with low population density. These areas are now almost inaccessible because the condition of the roads is so poor most of the year. To put the extent of this undertaking in perspective, one must note that the current road network has approximately 7,000 kilometers that are paved. The main justification for paving these roads is to provide agro-industrial grain producers in Brazil's Center-West region with a way to reduce transportation costs for their grains.

Traditionally, grain producers have brought their product to ports in the South/Southeast using trucks on unreliable roads. In 1997 a port facility in Rondônia (Porto Velho) was completed allowing grains to be transported from the Center-West to Porto Velho on highway BR-364 and then shipped to Manaus, with a reported savings in transportation costs of 20 percent. If the infrastructure improvements in the *Avança Brasil* program go forward as planned, producers may have several options, with even greater savings, ranging from rail transportation to Santos (Ferronorte Project) to transport on the Santarém–Cuiabá road, once it is paved, to shipping along the Amazon River to Belem.

Besides improvements in roads in the interior of the country, Brazil has also pursued the expansion of paved roads to borders with neighboring countries in the Amazon Basin. This is aimed at increasing economic integration between countries in the Basin. It is exemplified by the paving of the Manaus-Boa Vista highway (BR-174) connecting to Venezuela, BR-401 connecting Boa Vista to the Republic of Guiana, and BR-156 connecting Amapá and French Guiana. The incentives that shape current land use patterns in the area may therefore undergo considerable shifts.

To the extent that the projects mainly involve paving existing roads, rather than opening new roads in virgin areas, it seems reasonable to set aside spatial considerations in this analysis and to just look at the average cost reduction for transportation. However, one must consider that the results from these scenarios cannot tell us where the deforestation is going to occur. In all scenarios analyzed, a reduction in costs for transportation between the Amazon and the rest of Brazil increases deforestation rates (balanced plan closure). A 20 percent reduction in transportation costs for all agricultural products from the Amazon increases deforestation by approximately 15 percent in the short run and by 40 percent in the long run (Figure 5.3). Results of the model for transportation cost reduction are presented in Appendix E.

In terms of hectares deforested per year, reducing transportation costs would lead in the long run to approximately an 8,000 square kilometer increase in the annual deforestation rate. The reason for such a dramatic increase in the deforestation rate as

Figure 5.3 Change in deforestation rates if infrastructure to Amazon is improved

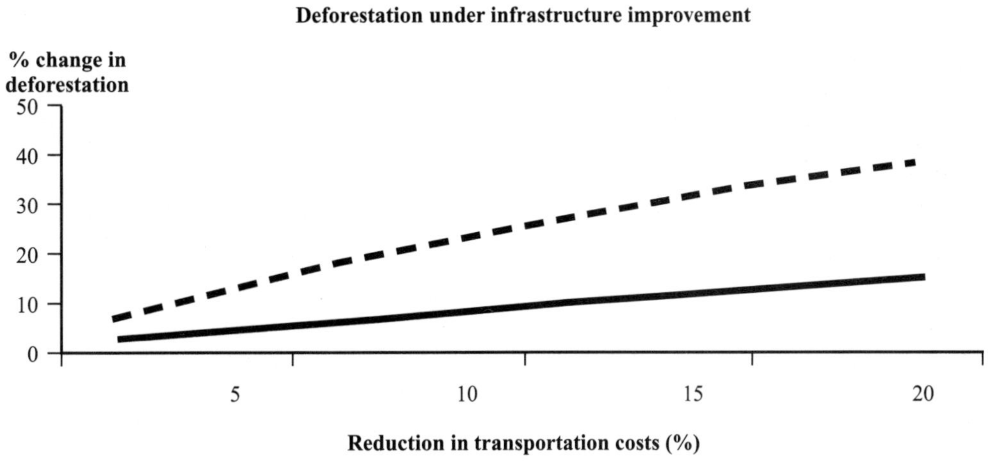

Deforestation under infrastructure improvement

% change in deforestation

Reduction in transportation costs (%)

transportation costs decrease is that transportation is a major cost component of producing agricultural goods in the Amazon. Therefore, infrastructure improvements have a large impact on profitability in agriculture. As agricultural production in the Amazon becomes more profitable, the price of arable land increases, thereby increasing the incentive to deforest. A 20 percent reduction in transportation costs leads to a 24 percent increase in the return to arable land in the short run and a 92 percent increase in the long run.

The increase in profitability leads, in the long run (with mobile agricultural labor and capital), to a 24 percent increase in production by smallholders and a 9 percent increase in production by large farms. However, welfare effects at the national level are very limited (rural households at the national level gain 0.6–0.9 percent in real income. This is because the increase in Amazonian production, except for the share that is exported, replaces previous production from other regions; therefore, the positive regional impact on development in the Amazon is offset by the negative impact on other agricultural areas of Brazil (Appendix E, Tables E.1–E.6).

The results of the analysis indicate a cumulative increase in deforestation of 160,000–240,000 square kilometers over the next two or three decades, for a 20 percent reduction in transportation costs. This result is in the same range (120,000–270,000 square kilometers) as that reported by Carvalho et al. (2001) for the impact on deforestation of the current government programs to expand infrastructure in the Amazon.

The simulations presented here could be further refined. It is assumed that transportation costs decrease for all Amazonian agricultural products; in fact, different products are affected differently by infrastructure improvements. An example of this effect is the port being built in Rondônia, which will reduce costs for grains but not for other products (at least initially). Another assumption is that smallholders, in the long run, will have perfect access to agricultural capital (and are on the same footing as large farms in terms of access to credit), which accounts for the good performance of smallholders in these results. In reality, smallholders may be constrained in their access to credit, which combined with the increases in land prices (implied by the higher returns), may put them at a disadvantage compared with the large farms. Having tried these simulations, however, the overall order of magnitude of a reduction in transportation costs on deforestation, whatever form it may take, remains unchanged. If the reduction in transportation costs is more precisely targeted to specific activities, the cost reduction will be greater than the 20 percent that is assumed across the board. But if one subset of producers is constrained from taking advantage of the cost reduction, another subset will.

Tenure Regime Regulation as the Means to Avoid Speculative Deforestation

In an unprecedented survey, the Brazilian government has mapped the country's land ownership structure in order to locate, one by one, cases of fraud and forgery of land-ownership titles. All over the country, the Instituto Nacional de Colonização e Reforma Agrária (INCRA) called upon landowners with properties greater than 10,000 hectares to prove their claims; if they could not, the land in question would return to public ownership. The investigation involved controlling registry documents for 93 million hectares. The result was that about 40 million hectares, mostly in the Amazon basin, were found to be fraudulent, and the titles were cancelled. The right of ownership of 22 million hectares is still undetermined (the deadline for landowners to appeal was December 31, 2001).

Of the total area of the state of Amazonas (157 million hectares), 27 million hectares were held illegally; 11.4 million hectares in the state of Pará and 1.3 million hectares in Amapá were owned illegally (Brazil, Ministry of Agrarian Development 1999). The cases of illegal land appropriation were usually characterized by changes introduced in original tittles of possession or ownership to increase the area of properties. Illegal land appropriation normally happens with the collusion of officials of real-estate registration notary offices. After obtaining registration in a real-estate public notary office, forgers repeat the same procedure at the State Land Institute, at INCRA's Registry Office, and at the relevant office of the Inland Revenue. Their goal is to obtain cross-registrations that support their fraudulent deeds and provide an appearance of legitimacy.

Fraud has historically been facilitated by institutional loopholes, such as the nonexistence of a single registry. Land-related organizations at the three levels of government (federal, state, and municipality) are not connected with one another. Contrary to what happens in other countries, in Brazil there are no special specific registration procedures for large areas. Data from the federal and state registries are not crosschecked and the federal registry, under the legislation now in force, is based upon the statements of owners. However, this is set to change soon: a law was passed in August 2001 that requires the development of a unified land registry system and more rigorous controls and penalties for land fraud.

The recent actions undertaken by the government on the combined front of investigating past frauds and avoiding future ones have the potential to truly change the land tenure mechanisms, as well as the perception of those considering fraud that they have a good chance of succeeding. This section analyzes the impact such changes may have on deforestation rates.

The economic literature linking deforestation to tenure regimes has either adopted a partial equilibrium approach (Mendelsohn 1994) or an econometric approach, based on the explanatory power of measures of tenure security using cross-country data (Deacon 1994; Alston, Libecap, and Schneider 1996; Deacon 1999). The approach adopted here is similar to Mendelsohn's partial equilibrium description; however, the context in this case is one of general equilibrium. Whereas, in the partial equilibrium setting, deforesters had the choice between sustainable forest uses and a destructive agricultural process with decaying physical output, in a general equilibrium framework deforesters have an array of additional choices. They can work for wages on large farms, or migrate to urban areas, or simply cultivate the already cleared land.

The assumptions made in simulating changes in tenure regimes have to be laid out. First, it is assumed here that deforestation is exclusively to clear land for agricultural purposes. Second, in the reference equilibrium, it is assumed that the returns to the deforestation activity are based both on acquiring property rights to unclaimed land and on future returns to agriculture.[37]

In the case of de facto property rights being acquired through deforestation, it is interesting to analyze the impact of a change in tenure regimes such that these property rights are made insecure through eviction. This change can be represented in one of two ways: as an increase in the discount rate equal to the probability of eviction (Mendelsohn 1994), or as a decrease in the expected time of residence on the plot before eviction. In the analysis that follows the latter of the two options is adopted.[38]

The results presented in Figure 5.5 show the percent change in the deforestation rate as a function of the expected time to eviction. The area between the two curves represents the domain of possibility described by a variability range in the discount rate of 15–30 percent, assumed here to bracket the true discount rate of farmers in the Amazon. The lower boundary is reached when the discount rate is 15 percent (low_disc). The upper boundary describes the impact if the discount rate is 30 percent (high_disc). The lower boundary shows a slow decrease in the deforestation rates, reducing the expected time of residence on the plot from 22 years to as low as 14 years (–7 percent) and decreasing more rapidly from there on (–12 percent for 12 years). The deforestation rate levels off at around 78 percent of its original value when the expected time of residence is reduced to 10 years.

The leveling off occurs because, as the risk of being evicted increases, it becomes more convenient to deforest previously tenured forest land than unclaimed land. The regime switches from deforesting as capitalization on acquisition of property rights (even if unsecured) to deforesting for the value added that comes from agricultural activities. Without considering global externalities, and given the 1994–96 average, the decrease in the deforestation rate is likely to be in the order of 5,000 square kilometers. This decrease, far from arresting deforestation, would still be a considerable improvement relative to the current trend, suggesting that the mode of tenure acquisition and enforcement should be top priority issues. If, on the other hand, the discount rate is higher than 15 percent, one can expect the leveling off point to be

[37]The net return to deforestation is different depending on whether land is titled or not; if the land is tenured one must subtract the returns from forested land in the computation. An average return is computed here by taking into consideration that about one-third of agricultural land in the Amazon has been reported to involve fraudulent titles.

[38]The difference between the two approaches is that the first assumes an equal probability of eviction over time, thereby additionally discounting for the risk at every point in time beginning at t=1. Instead the second option calculates the returns based on the expected time before eviction and therefore maintains the discount rate unchanged, but it does not include any revenues that could be obtained after the expected eviction date. The rationale underlying this choice is that the probability of eviction is unlikely to be constant over time. The two methods can be compared, if the probabilities are constant, by using the formula $E(T) = \frac{1}{p_{evic}}$.

Figure 5.4 Impact on the deforestation rate of regulating access to property rights

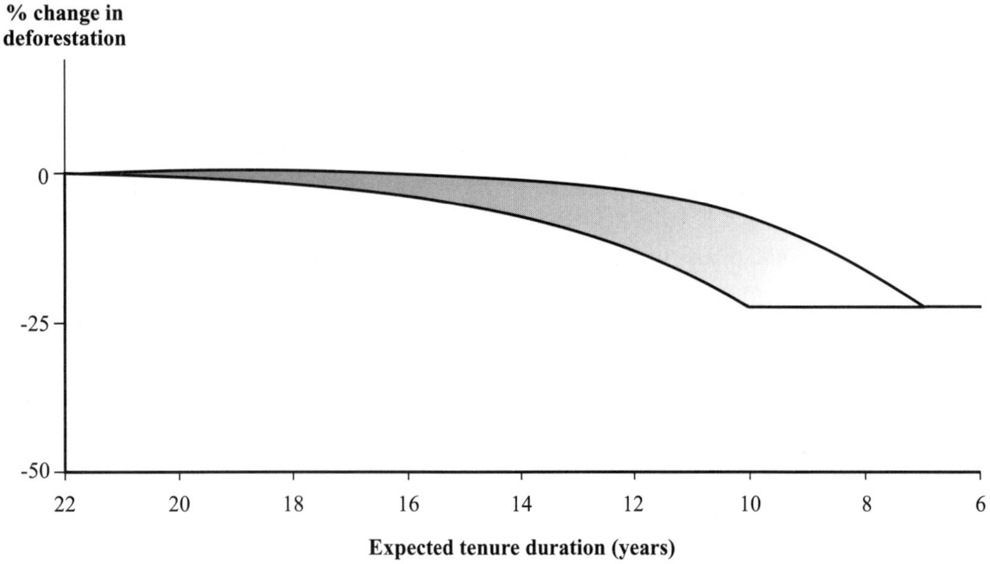

% change in deforestation

0
-25
-50

22 20 18 16 14 12 10 8 6

Expected tenure duration (years)

reached if the expected time to eviction is less than 10 years (the upper boundary, using a discount rate of 30 percent, reaches the leveling-off value in 7 years).

Previous results (Cattaneo 2001) assumed that all current deforestation occurs on unclaimed land, thereby causing results to overemphasize the impact of regulating tenure (a 63 percent reduction in deforestation rates). The simulations presented here take into account that only about one-third of the deforestation appears to be occurring on illegally claimed land. What is observed then is that if a share of the deforestation is already occurring on tenured land, this raises the "floor" on the deforestation rate because this component will not be affected by changing tenure regimes. Since data are not available on exactly how much deforestation is occurring on land without proper title, the results presented here (as those of Cattaneo 2001) are meant to capture the relative importance of tenure regime specification in determining deforestation rates.

Since, by construction, the analysis begins from an equilibrium point, the hypotheses in the literature that tenure leads to more deforestation can neither be validated nor contradicted (Vosti, Witcover, and Carpentier 2002). Nor can it be said that it leads to less deforestation (Deacon 1999). All that can be said is that relative to the 1995 base structure of the economy, assumed to be an equilibrium point, if unclaimed land is being deforested, then increasing the probability of eviction will decrease the deforestation rate to the point where it is profitable to clear only previously tenured land. In this respect, the results contradict the partial equilibrium results of Mendelsohn (1994), stating that the possibility of eviction leads to destructive land uses.

What Is the Impact on Deforestation of Technological Change in Agriculture?

After looking at the effects on deforestation of the exchange rate crisis, the improvement in infrastructure, and the tenure regimes, the study proceeds to analyze the impact of technological innovations in agriculture. Can technological changes in Amazon agriculture counterbalance the trend toward increasing deforestation rates? If so, do factor intensities of the technological innovation matter in determining deforestation rates? Are there differences between short- and long-run effects of such innovations? Last but not least, what is the impact on deforestation in the Amazon of technological change occurring outside the Amazon?

As the numbers for regional total factor productivity (TFP) gains during the period 1985–95 testify, all regions experienced substantial productivity improvements during this time.

Region	Overall TFP change in agriculture (percent)
Amazon	29.6
Northeast	24.2
Center-West	54.2
South/Southeast	21.6
Brazil	26.1

These estimates were computed based on data provided in Gasques and Conceição (2000). In relative terms, the greatest overall technological change occurred in the Center-West region, followed by the Amazon, the Northeast, and the South/Southeast. Given the premise that substantial technological change in agriculture has occurred in all regions of Brazil, and will likely occur in the future, it is worthwhile to investigate in some detail what the impact of different incarnations of technological change might be on deforestation and income distribution.

Intraregional Effects of Technological Change Occurring in the Amazon

At the local level, much has been done to examine the effects of technological change in the Amazon. Different farming and cattle-raising systems have been analyzed, with particular attention paid to the different dimensions of the issue, such as profitability, credit requirements, sustainability, and other factors that determine the adoption of any one specific technology (Toniolo and Uhl 1995; Mattos and Uhl 1994; Almeida and Uhl 1995; Serrão and Homma 1993; White et al. 2001; Vosti et al. 2001). The approach taken in this section refers to the impact of technological change as it occurs at the Amazon Basin level; more specifically, it expresses a modification in the structure of the producing sector as a whole in the region.[39] Technological change is assumed exogenous and, although the values express a reasonable range of possible change, they are not based on case studies.

Simulations include technological change in production of annuals, perennials, and animal products. For each activity, different types of technological change are analyzed: in a reference run, TFP is increased up to 70 percent in 10 percent increments (disembodied technological change). The other simulations replicate the productivity increase of the TFP case by acting on the productivity of specific factors (embodied in technological change). In the factor-specific cases, the extent of the factor productivity increase is inversely proportional to the factor's value share in production (to replicate the TFP case). For

[39]Different levels of technological change at the sectoral level can be associated with either a technology shift or with the extent to which a technology is adopted. For example, if all producers adopt a technological innovation causing a 50 percent improvement in TFP, this is equivalent in the framework to half the producers adopting a technology, leading to a 100 percent improvement in TFP at the plot level (ignoring nonlinearities).

Table 5.2 Types of technological change

Abbreviation	Name	Comments
TFP	Total factor productivity increase	Disembodied technological change: not associated with any specific factor
LAB_INT	Labor productivity increase	Improves labor productivity; attracts labor
CAP_INT	Capital productivity increase	Improves capital productivity: attracts capital
LABCAP	Labor and capital productivity increase (land saving)	Replicates land intensification: needs less land to produce a unit of output
DG_LBK	Land saving with decreased land degradation (agronomically sustainable)	Land intensification also increases sustainability and reduces degradation rate by 10% at each step

comparison purposes, the TFP index associated with an instance of technological change is defined as the TFP increase used as the reference for the simulation.[40] The different types of technological change are compared across simulations by representing the results relative to the TFP indices. Table 5.2 shows the different types of technological change in the simulations:

The simulations are carried out for the short run (1–2 years), in which agricultural labor and capital are confined to their regions, and for the long run (5–8 years) by allowing these factors to migrate interregionally. Results are obtained for terms of trade for Amazon agriculture, factor rentals, deforestation rates, and value added associated with smallholdings and large farm enterprises. This value-added differentiation is a proxy for income distribution in the region. It also serves a second purpose: it may indicate which forms of technological change are more likely to be adopted by different producer types. Due to space limitations, only the short-run results for value added are presented. The underlying assumption is that migration is triggered after the initial adoption of a new technology; therefore, it is desirable to find out if a technology is profitable at low TFP indices and without migration. Value-added shares are good proxies for income distribution in the short run because migration is not allowed. Additional results for the technological change scenarios can be found in Appendix F, Tables F.1–F.3.

Improving Annual Crop Technology Inside the Amazon

In the short run, increasing the productivity in annuals cultivation may increase or decrease the deforestation rate depending on the type of technological change (Figure 5.5). The TFP case, in which factor productivity in annuals is increased by the same amount for all factors, appears to lead to the greatest deforestation, followed closely by capital-intensive technological change (CAP_INT). The reason these two forms of innovation have the strongest push toward deforestation is that arable land appreciates considerably as a consequence of the productivity improvement (Appendix F.1, Table F.1). For the TFP scenario, this effect is direct because land productivity improves; however, for the CAP_INT scenario, an even greater indirect appreciation of arable land is observed, as a result of derived demand for factor inputs in annuals (a very large improvement in capital productivity is required to replicate the TFP case

[40] A TFP index equal to 1 indicates an equivalent TFP increase of 10%.

Figure 5.5 Change in deforestation rates for technological change in annuals production

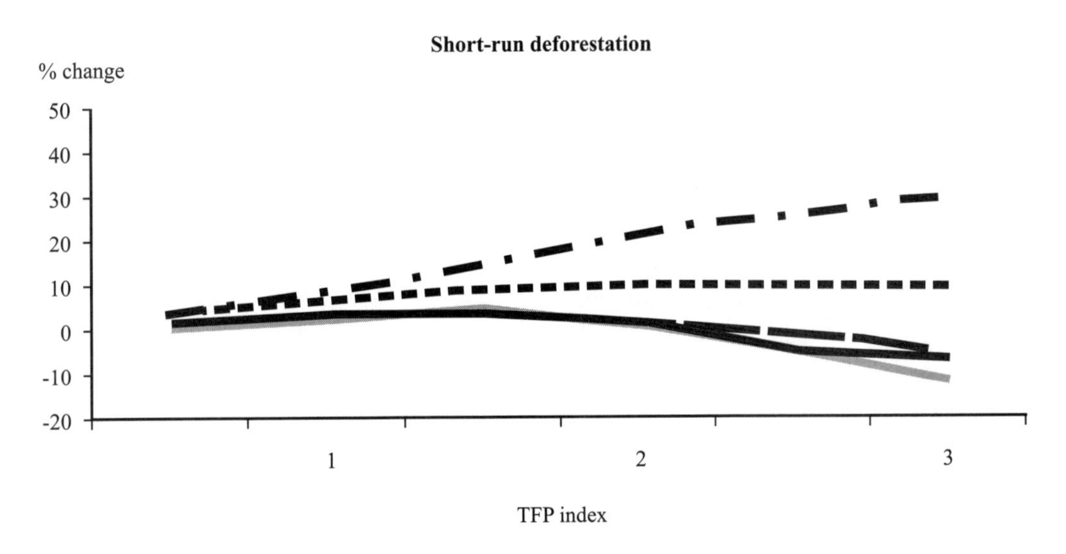

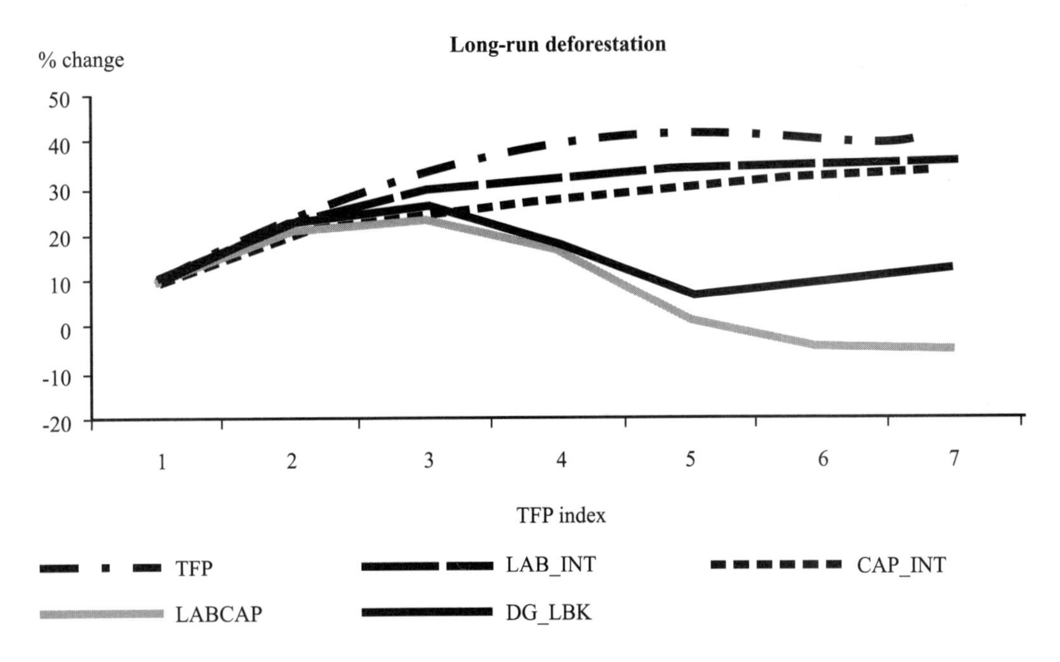

Notes: The full names for the types of technological change abbreviated in these figures are given in Table 5.2.

because the capital-intensity of annuals in the Amazon is low).

The price of pastureland, in the short run, acts as a dampening factor that proves to be important in the labor-intensive (LAB_INT) and land-saving (LABCAP) cases. In these two cases, arable land also appreciates considerably; however, the price of pasture decreases accordingly, making it less attractive to deforest since the newly deforested land will soon be transformed to grassland.

Increasing the productivity of labor in annuals is equivalent to an increase in the endowment of labor in the sector, and consequently resources are attracted to annuals. This result implies that there is a lower demand for pastureland as a factor, since it cannot be used in the production of annuals. This effect can reduce the incentive to deforest to the point of slowing deforestation; however, this occurs only after a certain threshold of technological innovation is adopted by the producers. In the simulation presented here, this threshold is 20 percent in TFP terms (TFP index = 2).

In the long run, allowing for migration of labor and capital between regions, the results change drastically (Figure 5.5). Technological improvement in annuals production leads to higher deforestation rates, unless highly land-intensive technologies are widely adopted. Even if such a case were to materialize, in the early stages of adoption (TFP index 1–3), deforestation would increase. The LAB_INT scenario is particularly interesting given that it appeared to be very promising in the short run. The difference is that in the long run the annuals sector attracts labor and capital from outside the Amazon, and arable land becomes a scarce resource. This causes a large increase in the value of arable land. Furthermore, the dampening factor of the decreasing pasture prices is reduced because resources no longer have to be diverted only from other Amazon agricultural activities but from the Brazilian economy in general.

The land-intensive scenario (LABCAP) performs well, in deforestation terms, in the higher range of the TFP index. This can be attributed in part to the finite amount of rice, manioc, and beans that the national market can absorb from the Amazon: land is less of a constraining factor under this technology, and the greater increase in production that it makes possible causes terms of trade to deteriorate. This feedback mechanism causes a reduction in migration flows relative to the labor-intensification case. The adjustment outside the Amazon to this greater production in annuals has an impact on the terms of trade for livestock produced in the Amazon, lowering the return to pastureland and thereby helping reduce the incentive to deforest.

The impact of improved sustainability of annuals (DG_LBK) combined with intensified land use proves to be interesting in the long run. There are two countervailing processes linked to sustainability: the first is a stock effect, whereby less degradation means a greater stock of arable land, which reduces the demand for deforestation. The second process depends on producers' expectations about the revenue flows to be had from arable land: if agriculture is more sustainable, high revenues from annual cultivation can be obtained for a longer period of time, increasing the demand for arable land. In the simulation presented here the stock effect is minimal: when a TFP index is more than 4, the expectation effect clearly dominates (as can be observed by comparing DG_LBK to LABCAP).

Given that annuals production is labor-intensive, improving labor productivity is clearly a welfare-increasing option, particularly for smallholder agriculture (Figure 5.6). In fact, it is the only option among the possible changes in annuals technology that improves conditions for smallholders. This occurs because capital markets are segmented in the short run. Smallholders cannot implement any technology requiring more capital because they do not have access to it. Therefore, large farm enterprises

Figure 5.6 Short-run change in value added in the Amazon region from technological change in annuals production, large and small farms

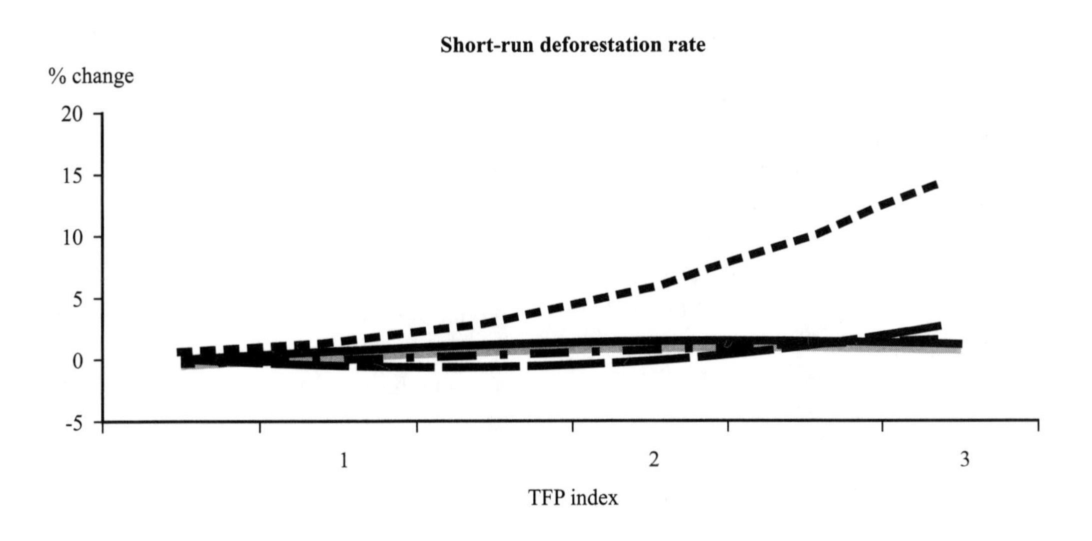

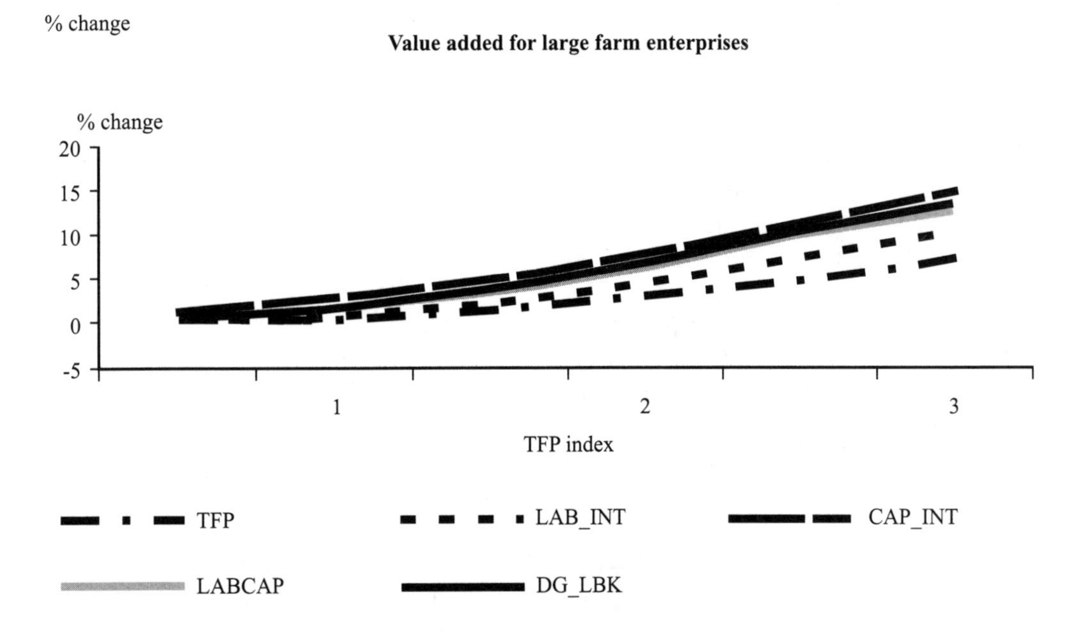

Notes: The full names for the types of technological change abbreviated in these figures are given in Table 5.2.

having access to capital stand to gain from improvements involving the use of capital (CAP_INT, LABCAP, and DG_LBK), while smallholders do not.

Labor-intensive technologies also considerably improve large farms' value added (they can hire off-farm labor), but the best option for farm enterprises is land-intensive innovation (LABCAP and DG_LBK in Figure 5.6). This occurs because their production of annuals balances labor and capital, thereby reducing the diversion of capital to the livestock sector observed for smallholdings.

Improving Perennials Technology Inside the Amazon

In both the short and the long run, increasing productivity in perennials cultivation has a generally positive potential for reducing the deforestation rate (Figure 5.7). In the short run, any factor- embodied technical change has the effect of lowering the price of pasture and the counterintuitive effect of decreasing the demand for arable land (the lower bound for arable land rent becomes binding), letting arable land be used as pasture. The underlying cause of this shift is that perennials make intensive use of labor and capital per hectare cultivated (more than annuals). This result implies that as resources are drawn to perennials, there will be less overall demand for arable land. A second reason for the decrease in deforestation is that perennials, as opposed to annuals, do not cause transformation of arable land to grassland; therefore, there is a stock effect whereby the amount of available arable land increases, tending to reduce the demand for deforestation.

In the short run, where the factor productivity in perennials is increased by the same amount for all factors, the TFP appears to have almost no impact on deforestation because the reduction in demand for arable land is offset by the increase in land productivity, which raises the return to arable land. All factor-specific improvements lead to a substantial decrease in the deforestation rate. Given the different deforestation rates associated with the land-saving and TFP scenarios, it is important to distinguish in practical terms the difference between these two forms of innovation. The land-intensive case for perennials assumes that each unit of capital and labor has become more productive: for example, a variety of coffee that allows more trees to be planted on a hectare would be more productive, but more capital and labor inputs would be required for it to be successful. The increased demand for these factors comes at only a slightly lower cost than the revenue increase from the productivity gain. The TFP case assumes that the improvement is not exhausted by the increased payments to labor and capital, and therefore, a share of the value of the increased production is associated with land. An example of a TFP improvement might be a technology that makes each tree more productive but maintains the same tree density. Labor and capital costs associated with planting this new variety may be higher per tree, but the number of trees is unchanged: an extra surplus, to put it in Ricardian terms, is associated with the productive possibilities of land.

In practical terms, the chances are that a technological improvement in perennials will always have some spillover to the value of land. In any case, as long as the improvement in the productivity of land does not exceed the improvement in the productivity of the other factors, deforestation will decrease in the short run.

In the long run, the results are still encouraging for perennials. However, more care needs to be taken in determining the form of technological change to be adopted. In the long run, the labor-intensive innovation brings further improvement in deforestation, because, with migration, even more substitution of production of annuals for perennials may be carried out. The land-intensification scenario changes slightly from the short run to the long run. Although the underlying process is unchanged, with migration, there is no surplus arable land to be used as pasture; in fact, arable land increases in value. However, deforestation is still reduced due to the dampening effect of lower returns to pasture land, which occurs as factors shift toward the production of perennials. This dampening effect is also present in the TFP and the capital-intensive scenarios; however, it is not enough to offset the prospect of higher returns from

Figure 5.7 Short- and long-run changes in deforestation rates from technological change in perennials production

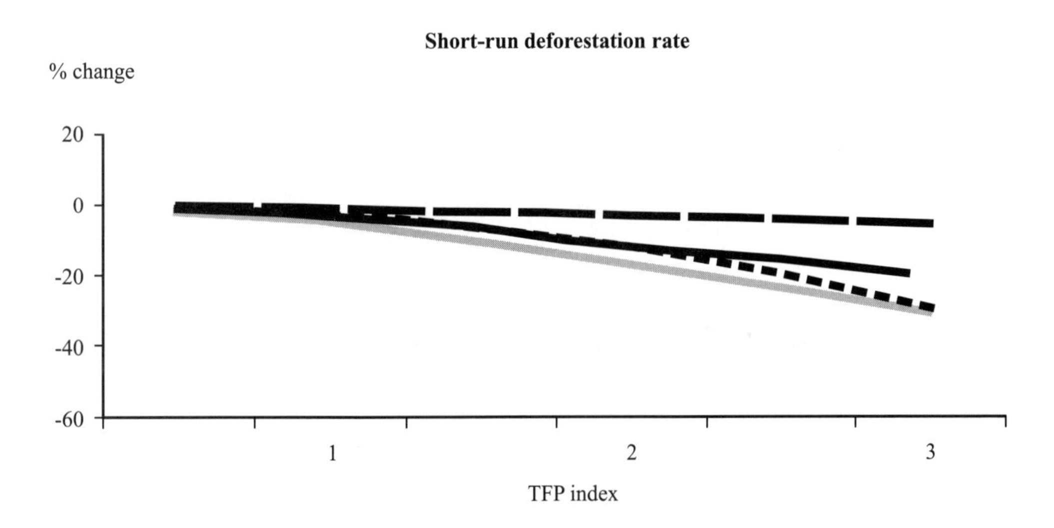

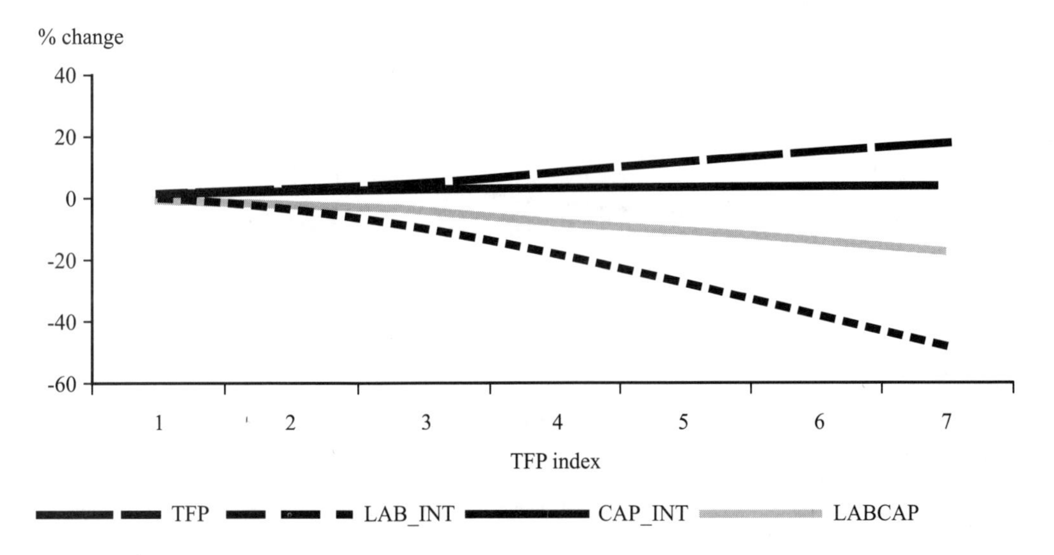

Notes: The full names for the types of technological change abbreviated in these figures are given in Table 5.2.

arable land. Therefore, deforestation increases in the long run in these two cases.

In the short run, small farms appear to gain relatively more income than large farm enterprises when production is shifted to perennials (Figure 5.8). This result arises, in part, from the fact that smallholders are already producing the majority of perennials in the Amazon ($620 million, compared with $130 million for large farms).

The implication, in this framework, is that fewer resources have to be diverted

Figure 5.8 Short-run change in value added in the Amazon region from technological change in perennials production, small and large farms

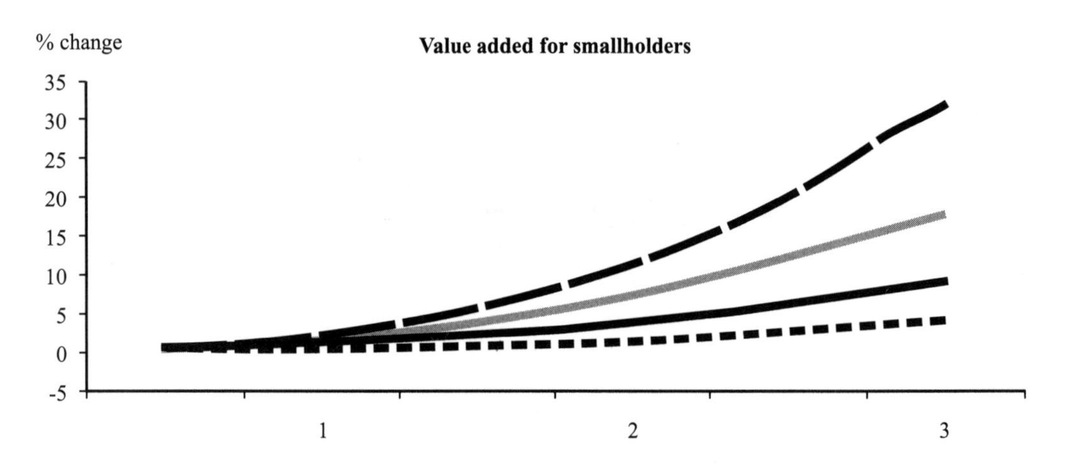

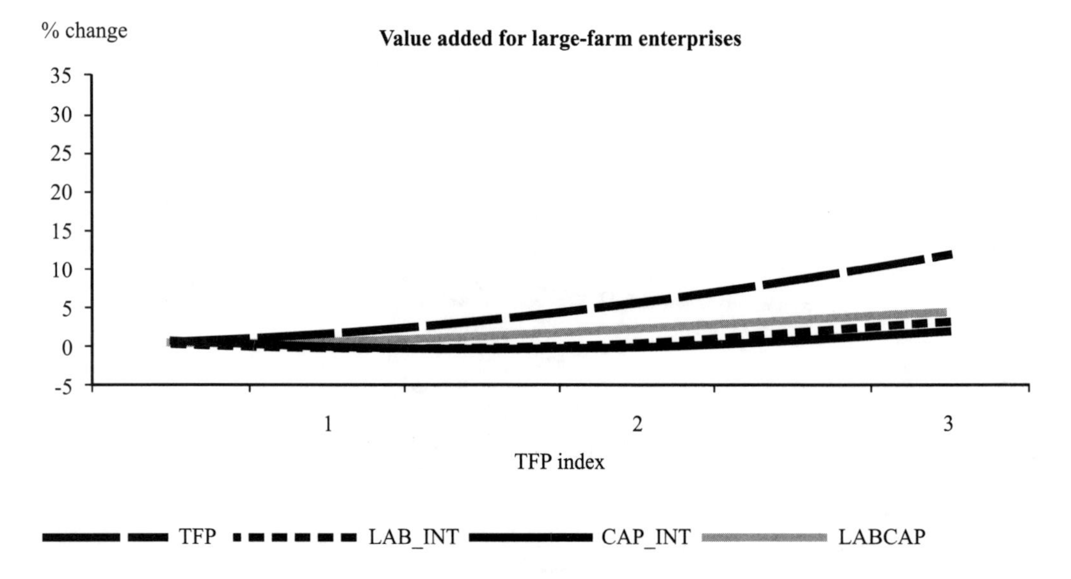

Notes: The full names for the types of technological change abbreviated in these figures are given in Table 5.2.

from other activities for this productivity push to be felt. This may actually not be the case in the real world because the smallholder capital in perennials consists mainly of trees, which in the case of technological change may have to be replaced for the productivity improvement to occur. The gain for smallholders may, therefore, be overstated in the results. To summarize, labor-intensive change appears to be the best option for smallholders as a whole because of their capital constraints. Conversely, capital-intensive technological change is the best option for large farm enterprises.

Figure 5.9 Short- and long-run changes in deforestation rates from technological change in animal production.

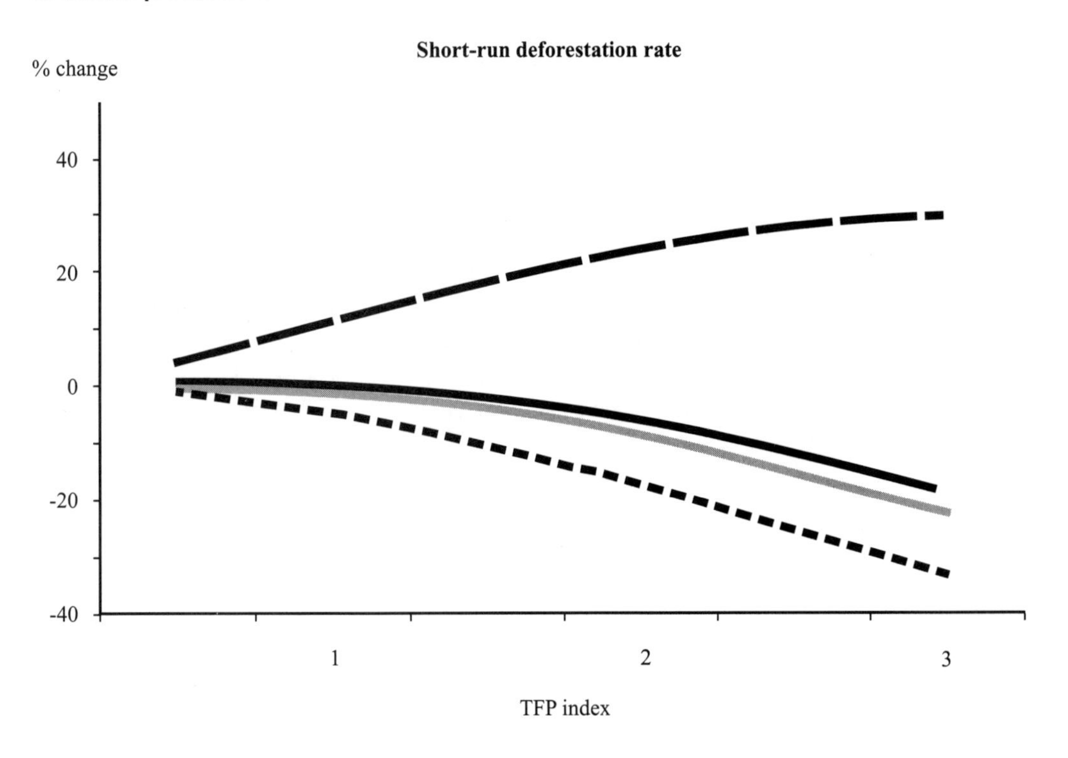

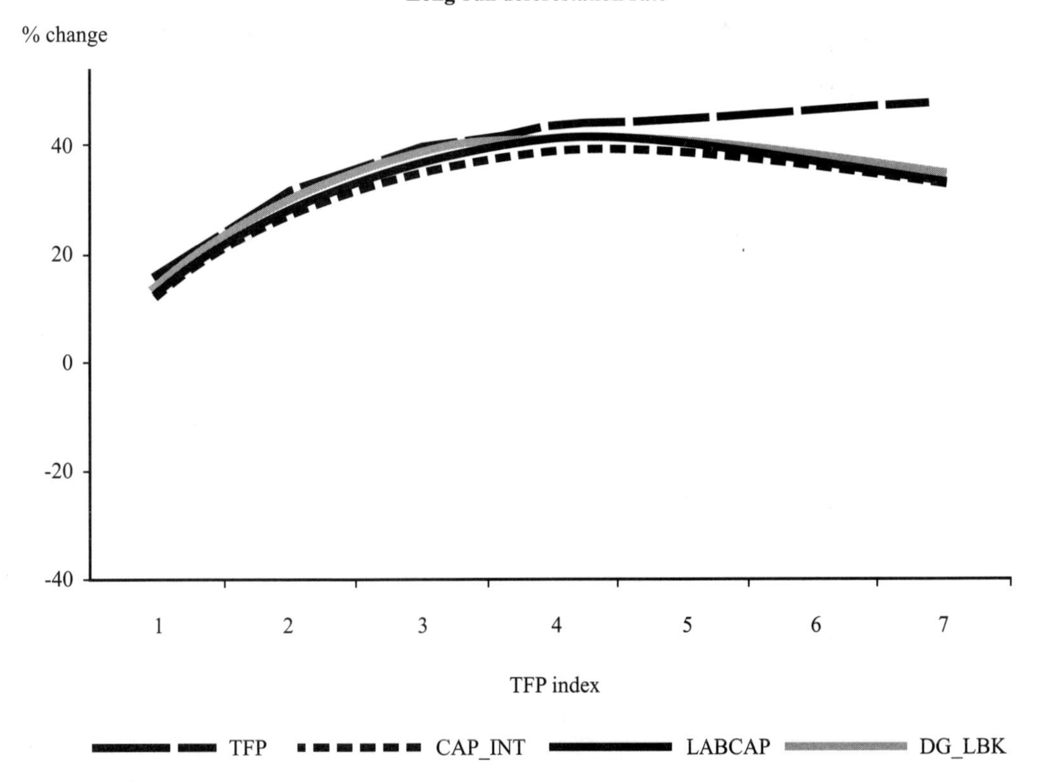

Notes: The full names for the types of technological change abbreviated in these figures are given in Table 5.2.

Improving Livestock Technology Inside the Amazon

There is an expectation that improved pastures in the Amazon, by allowing for a more land-intensive production system (combined with appropriate regional policies), will reduce deforestation (Mattos and Uhl 1994; Arima and Uhl 1997). These expectations appear to express a short-run view, in that they do not take into consideration the long-term effects of a more profitable ranching sector in the Amazon. In fact, in the short run, all improvements not directly affecting the productivity of land do lead to a reduction in deforestation, but this result does not hold true in the long run (Figure 5.9).

With no migration of labor or capital allowed, understanding what happens is straightforward: some arable land is used for pasture as the livestock sector becomes more profitable. This is the least-cost solution in the short run. In fact, with a TFP index equal to 3, demand for arable land is reduced by 70–80 percent in all scenarios except the TFP case. Here too the results may overstate reality, since farmers' food security constraints are not considered, and capital is assumed to be mobile for both large farms and smallholder farms. In reality, capital in the livestock sector is embodied by the herd, which has a natural growth rate that cannot be adjusted in the short run.

In the long run, as resources can be attracted from outside the Amazon, the increased demand for pasture is met by increasing deforestation. The surprising result is that not only does the return to pasture land increase substantially, but the price of arable land also increases. The increased price of arable land comes about because production of annuals leads to land degradation and subsequent use of the land as pasture; as keeping the land in pasture becomes more attractive, the demand for arable land increases in expectation that it can be used as pasture in the future. In fact, in all the long-run scenarios, production of annuals increases alongside that of livestock (although at a lower rate). Although perennials are also produced on arable land, they do not cause degradation, do not expand, and may actually be reduced. In all scenarios, *improving livestock productivity in any way will substantially increase deforestation in the long run.*

From a farmer's perspective, improvement in livestock technologies is a top priority. The returns from capital-intensive technological innovation or from land intensification in livestock would be extremely high for all producers in the Amazon, compared with improvements in annuals or perennials (Figure 5.10). Returns from TFP improvements would also be significant but less pronounced.

To return to a familiar theme, improving the productivity of the intensive factor for an activity is bound to reinforce the expansion of the activity. Furthermore, in a regional economy like the Amazon, where labor scarcity is a major constraining factor, livestock is a very attractive option, and in fact, it is already well established as a productive activity. This is reflected by the wage change for unskilled labor associated with technological change (+9 percent for TFP index = 3 in LABCAP for livestock, compared with 47 percent for the same type of change in annuals, as shown in Appendix F, Table F.1).

To conclude this section, the impact of different types of technological change in the Brazilian Amazon are summarized and compared in order to determine if there is a trade-off between developing agriculture and reducing deforestation. The best option, in deforestation terms, is technological change in perennials, which may also have positive effects on income distribution by favoring smallholdings. However, from a purely revenue-driven perspective, livestock is the best alternative for both small and large farms. This leads to an unfortunate dilemma because any form of technological improvement in livestock will lead to greater deforestation rates in the long run. Improvement in production of annuals,

Figure 5.10 Short-run change in value added in the Amazon region from technological change in animal production.

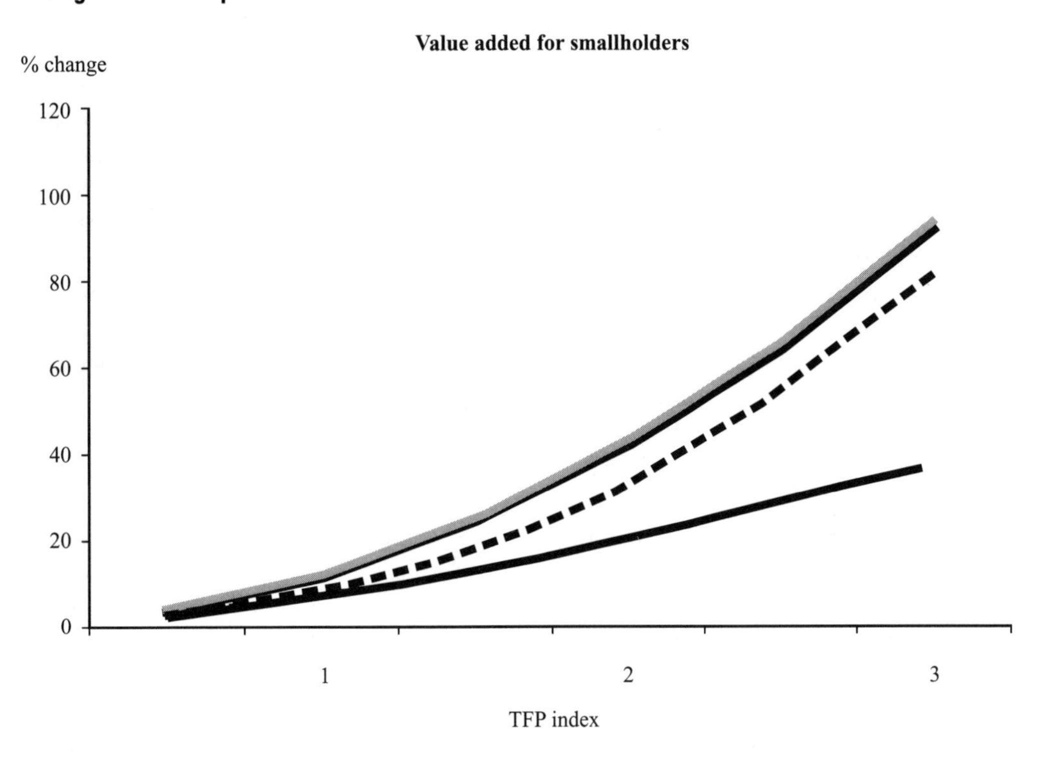

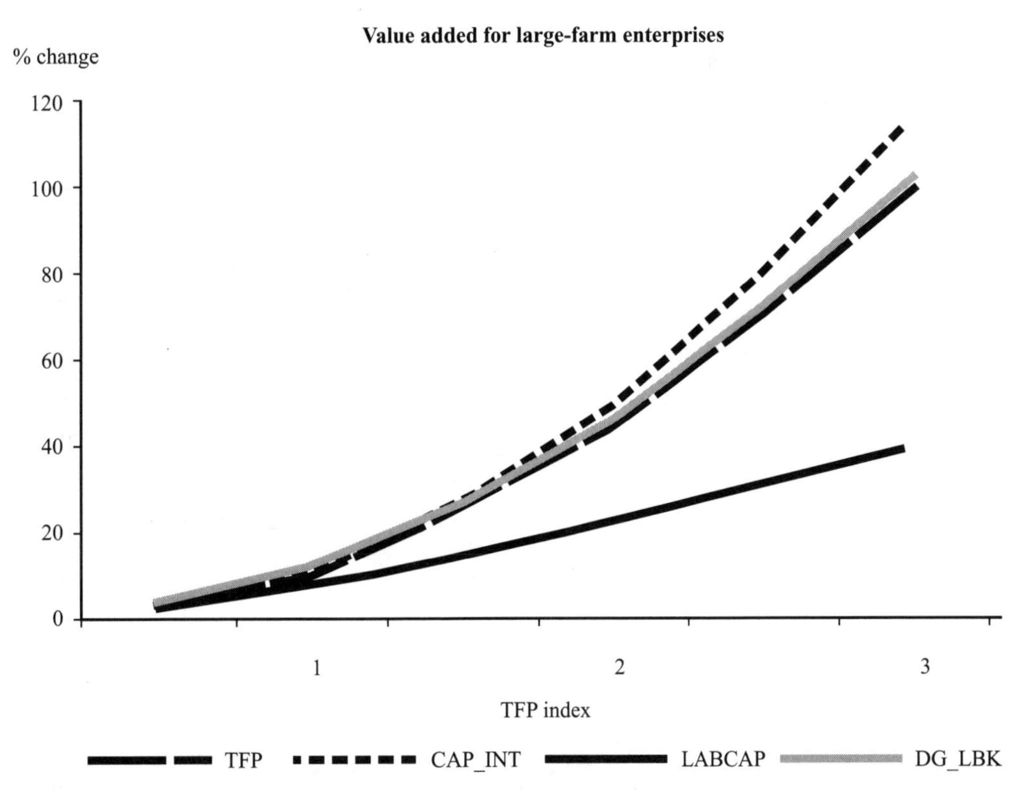

Notes: The full names for the types of technological change abbreviated in these figures are given in Table 5.2.

while possible in certain parts of the Amazon, would probably bring higher deforestation rates with returns of about the same magnitude as perennials; therefore, improvement in annuals does not seem particularly appealing.

Care must be taken in analyzing the effects of the possible technological innovations available in any specific activity. There are differences in short- versus long-run effects and also among the different factor-specific productivity changes. In the short run, TFP scenarios always lead to the greatest deforestation rates. However, this result does not carry over to the long run for annuals and livestock, for which labor-intensive change leads to greater deforestation rates. Land intensification performs well in reducing deforestation in all scenarios, except livestock in the long run, in which case it leads to the highest deforestation rates observed in the simulations.[41]

Interregional Effects of Technological Change Outside the Amazon

The policy relevance of technological change occurring outside of the Amazon lies in the past as much as in the future. Past modifications in agricultural technology in the South/Southeast region of Brazil are thought to have led to migration in the 1960s and 1970s and therefore played a role in the opening of the Amazon frontier. In the 1980s and 1990s technological improvements in agricultural production in the Center-West, particularly in soy production, reportedly displaced livestock producers from that region. The livestock producers decided to sell their land to soy producers and move their livestock operations to frontier areas (Schneider 1992). Future technological change in agriculture will likely have an impact on deforestation and income generation.

This section tries to interpret the past and, drawing on this first step, analyzes the implications of ongoing and future technological change. Thus it is divided into two components: the first analyzes retrospectively the impact on deforestation and income of agricultural technological change that occurred outside the Amazon during 1985–95. The second part, analyzing the impact of different types of possible technological change, focuses on each type of change separately rather than analyzing the overall effect as in the first step. The rationale behind this approach is to first evaluate the statement that the technological change occurring during the period was a factor in increasing deforestation in the Brazilian

Table 5.3 Shift from natural to planted pasture, by region, 1985–1995

	Area in 1985 (million hectares)		Change in pasture area 1985–95 (%)	
Region	Natural	Planted	Natural	Planted
Amazon	11.8	9.1	–18.1	61.8
Northeast	23.3	11.9	–14.2	2.0
Center-West	29.0	28.0	–39.8	34.7
South/Southeast	46.2	22.8	–32.9	20.6
Brazil	110.2	74.0	–29.2	34.6

Source: IBGE 1998a.

[41]The results for technological change in the Amazon presented in this section are consistent with those in Vosti, Witcover, and Carpentier (2002) for the Western Brazilian Amazon. In both studies technological innovation in livestock has a positive effect on income but increases deforestation, while perennials are good for small-farm income and for reducing deforestation.

Amazon. Having qualified such a statement, the analysis shifts toward the isolated impact of different types of technological change that may occur in the future.

A Retrospective Scenario: Impact of Productivity Improvements Outside the Amazon, 1985–95

To undertake the first step, a scenario was constructed based on measures of agricultural technological change reported in the literature for the period under consideration. The scenario relies on data from Gasques and Conceição (2000) for TFP in agriculture at the state level, Evenson and Avila (1995) for TFP changes in annual crops in selected states, and the 1995/96 Agricultural Census (IBGE 1998a) for shifts from natural pasture to planted pasture as a proxy for technological change in livestock activities.

To obtain the productivity changes with the regional specification adopted in the model, the estimates at the state level had to be aggregated to the regional level by weighting the productivity change according to the states' share of agricultural land in their respective regions. This resulted in the estimates reported in Table 5.3. What emerges from the aggregated estimates is that, except for the Northeast, all areas had large increases in the area in planted pasture (mostly substituting natural pasture), indicating substantial technological innovation among livestock technologies.[42]

The data on annuals shown in Table 5.4 summarizes Evenson and Avila's (1995) findings by presenting the range of productivity improvements in terms of lower and upper estimates of productivity improvement for a set of annual crops (in Evenson and Avila, different states in region had different levels of innovation occurring). Although the South/Southeast region had noticeable increases in productivity of annuals, the Center-West, with the exception of wheat, has had greater technological improvement and consistently higher productivity. This is probably due to the improvements in adapting these crops to climatic and soil conditions in the cerrado, a type of savannah primarily found in the Center-West.

To conclude the data used in the construction of the retrospective scenario, the estimates of regional improvements of TFP in agriculture as a whole are computed from the data provided in Gasques and Conceição (2000) for the 1985–95 period. In relative terms, the greatest overall technological change occurred in the Center-West (54 percent), followed by the Amazon (30 percent), the Northeast (24 percent), and the South/Southeast (22 percent). These data, along with the rest of the information in Tables 5.3 and 5.4, were used to construct the retrospective scenario presented in Table 5.5. The scenario was conceived to capture the relative changes in technology that occurred from 1985 to 1995, and not as an accurate representation in absolute terms. The improvements reported in Table 5.5 were obtained by considering what the overall regional TFP improvements in agriculture were, and in what activities these were likely to be occurring. What the scenario captures, in relative terms, is (1) the great improvement in the production of annuals among large farms in the Center-West and South/Southeast regions, (2) a considerable improvement in livestock productivity in all three regions, and (3) technological innovation in perennials production in the Northeast (as reported in an anecdotal manner in

[42]Gasques and Conceição (2000) also report increasing specialization in poultry in the Northeast and in poultry and swine in parts of the Center-West and South/Southeast (which would not be picked up by shifts to planted pasture even though they are livestock activities).

Gasques and Conceição 2000). In absolute terms, the magnitude of regional technological change obtained by aggregating the sectoral innovation (using a value-of-production weighted average of innovation rates), was made to match the regional technological change obtained based on Gasques and Conceição (2000).[43]

As expressed in the previous section on technological change within the Brazilian Amazon, innovation outside of the Amazon can also take on different forms in terms of what factors are affected. Although, simulations were run for the same combinations of technological change as in the previous section, the results presented here cover only the "balanced" TFP improvement and the combined labor and capital productivity improvement (LABCAP). Limiting the results to these two cases is sufficient in that these simulations are less sensitive to which factor's productivity is increased and more dependent on which activities are being affected and their impact on terms of trade for Amazonian products. In other words, because innovation in the Amazon creates a "pull" on factors, the form of technological change matters. But the impact of extra-Amazon innovation on the Amazon is mainly through its impact on activities because the "push" of any losing factor is diluted by the many paths that migration can take (with the path to the Amazon being simply one among several options).

What emerges quite clearly from the simulation, as one can observe in Figure 5.11, is that overall technological change outside the Amazon did not cause greater deforestation; in fact, it may have led to a reduction in deforestation rates of 15–35 percent. Since considerable uncertainty exists about what type of technological change occurred, whether small farms took part in these innovations, and the extent of innovations in the Northeast, Figure 5.11 presents a range of possibilities of what may have occurred in 1985–95. The results indicate that accounting for all the productivity changes in Table 5.5 (considered here to be the "historical scenario") would have led to a 27–35 percent decrease in the deforestation rates. This reduction is less pronounced if the Northeast or small farms or both are excluded from the innovation that took place during the period; with these limitations, deforestation rates would be reduced 15–27 percent, depending on the type of technological change.

Letting aside the uncertainty on the magnitude and form of technological innovation that occurred, it appears that innovation in the Northeast, among small farms, and even among the large farms in the Centerwest and South/southeast contributed to limiting deforestation. If innovation in the Northeast is taken away in Figure 5.11, deforestation in the Amazon decreases by 24 percent instead of 27 percent. The reason a lack of technological improvement in the Northeast leads to a smaller decrease in the deforestation rate is that some capital is moved out of the Northeast and into large-farm livestock production in the Amazon.

Table 5.4 Increase in annual yields in two regions 1985–1995 (%)

Crop	Center-West	South/Southeast
Maize	55–67	15–33
Beans	20–48	–13–40
Rice	44–85	29–61
Wheat	35–63	53–98
Soybeans	28–31	7–27

Sources: Evenson and Avila 1995.

[43]The numbers reported in Table 5.5 were tested for robustness using Monte Carlo simulations. Ten thousand simulations of different types of technological change consistent with the aggregate regional numbers obtained from Gasques and Conceição (2000). The results obtained in the Monte Carlo simulation are consistent with those presented in this report; in fact, the results shown here represent the mean of the outcome in the Monte Carlo simulations.

Figure 5.11 Change in deforestation rates for non-Amazon technological change: Scenario replicating innovation in 1985–95

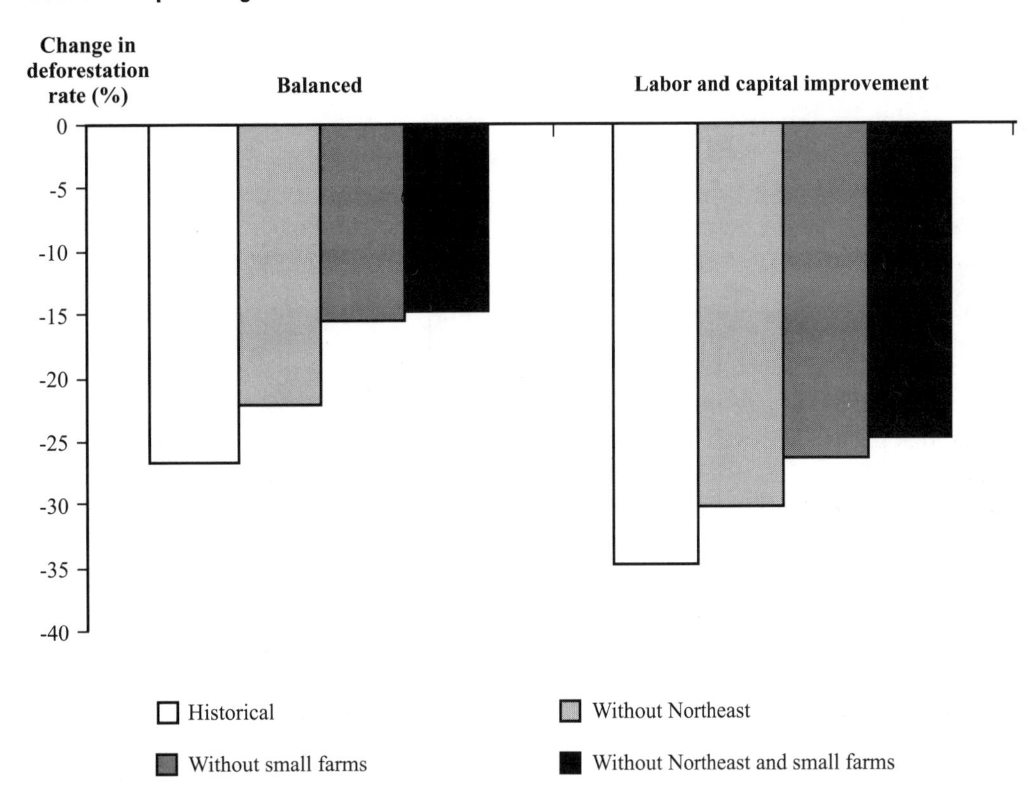

Small farm innovation contributed to limit deforestation by avoiding relocation of small farms to the agricultural frontier; therefore, when small farm innovation is removed the decrease in deforestation rates is just 15 percent.

Table 5.5 Replicating productivity improvements for the period 1985–1995: a retrospective scenario (% change)

Land use	Northeast	Center-West	South/Southeast
Small farm			
Annuals	20	24	21
Perennials	40	40	11
Animal production	26	26	21
Large farm			
Annuals	20	63	36
Perennials	40	18	11
Animal production	20	52	21

Note: Numbers represent percent change over the whole period. These can be converted to annual rates of change. For example, a 63 percent improvement in annuals in the Center-West is equivalent to an annual rate of change of 5 percent.

Technological change outside the Amazon appears to limit deforestation independently of whether large or small farms innovate, and where the innovation occurs outside the Amazon. However, the production technology being innovated, whether livestock, annuals, or perennials, is a determining factor of the impact innovation has on deforestation. At first glance it would appear from the results in Figure 5.11 that the general statement that "technological improvement in Brazilian agriculture caused movement to the agricultural frontier" is incorrect; however, the qualified statement that improvement in annuals—soy in particular—caused deforestation is correct (Kaimowitz and Smith, 1999). This can be seen in Figure 5.12, where the same simulations were performed without improvement in livestock technologies. The

Figure 5.12 Change in deforestation rates for non-Amazon technological change: What if innovation in livestock had not occurred?

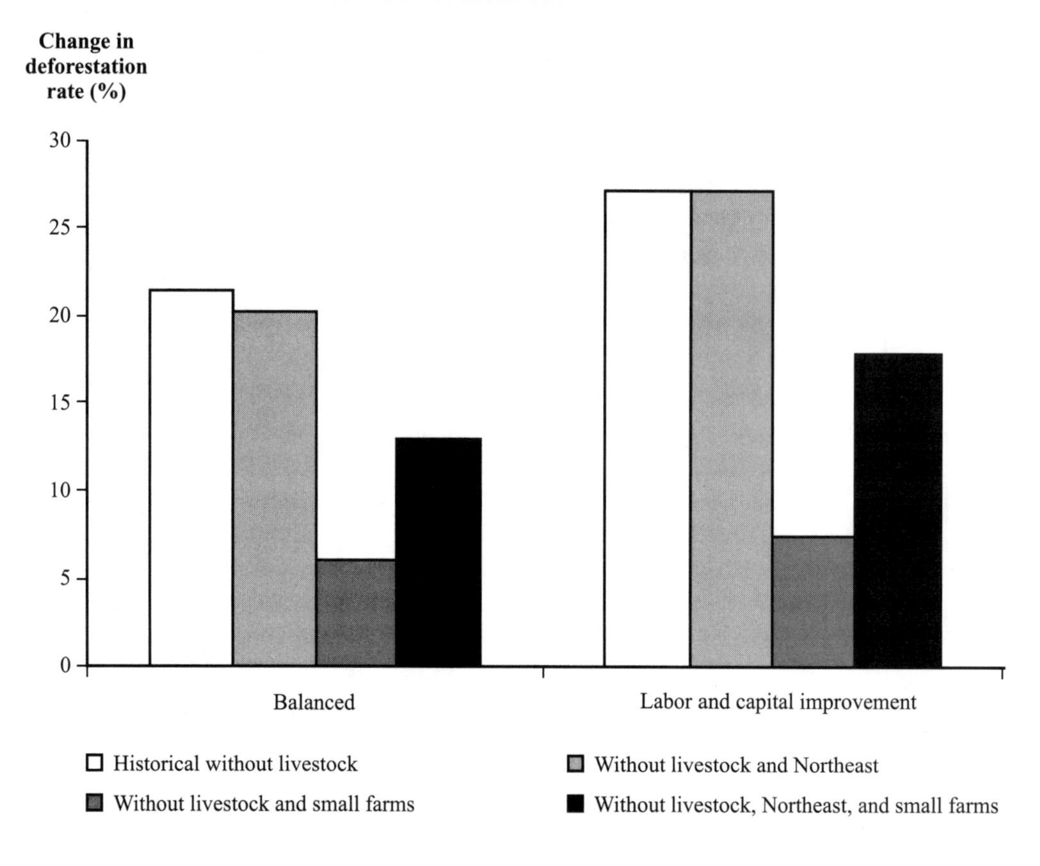

deforestation results are reversed: technological improvement in annuals, if not complemented by innovation in livestock, causes major increases in deforestation rates.

Taking away livestock innovation from the historical scenario leads to an increase in deforestation of 22 percent to 27 percent depending on what form the technological change takes. This highlights the finding that terms-of-trade effects favor production of livestock in the Amazon when production in other regions shifts toward annuals and perennials as a result of technological improvement. Technological innovation in the Northeast, if livestock technology innovation had not occurred in Brazil, does not play a significant role in determining deforestation rates. This is implied by the very small reduction in deforestation rates in Figure 5.12 when Northeast technological change is removed. This limited role is due to the fact that removing innovation in the Northeast has two counterbalancing effects: the first is to push resources toward other regions (Amazon included), while the second is to depress the Amazon terms of trade for livestock goods because Northeast resources will shift toward livestock production if no innovation occurs in the region. These two effects are similar in magnitude and effectively cancel out in determining deforestation in the Amazon.

Combining a lack of livestock improvements with the assumption that no innovation occurs among small farms has a considerable impact, limiting the increase in the deforestation rate to between 6 and 8 percent. This phenomenon occurs mainly because there is a substantial share of livestock being produced on small farms (especially in the South/Southeast, see

Table 4.6). Therefore, if technological improvements occur on small farms (but not in livestock activities), a substantial amount of resources will be drawn away from livestock in those regions, thereby improving the terms of trade for Amazonian livestock. When these improvements are taken away, this diversion of resources no longer occurs, and so deforestation rates do not increase as much as in the "historical without livestock" scenario.

The final scenario (without livestock, Northeast, and small farms) which is a combination of the two scenarios above, is somewhere in between the two in terms of deforestation rates. This scenario predictably balances the two effects found in the previous scenarios. The lack of innovation in the Northeast causes capital to migrate to the Amazon; however, the terms of trade for Amazon livestock are not as good when there is no technological improvement among small farms outside the Amazon, and none in the Northeast in general.

It is clear from these results that the process of technological change in agriculture could potentially affect deforestation in drastically different ways. It is interesting to analyze whether this also applies to agricultural income generation. The effects on income of such widespread technological change in agriculture will occur mainly through terms of trade for agricultural goods produced in the different regions. If production of one good is greatly increased in one area of Brazil, it will likely replace production from another region through price repercussions. For this reason, the overall impact of innovation on income generation at the national level is less dramatic than what happens at the regional level (Table 5.6).

Looking first at the historical scenario, one observes that technological change outside the Amazon has a negative impact on agricultural income generation in the Brazilian Amazon (Table 5.6). This result is not surprising, considering that in this scenario other regions are becoming more efficient, while the Amazon is assumed not to innovate. The decrease in Amazon agricultural value added is in the range of 22 to 26 percent. Large farms would be particularly hard hit if the Amazon were to lag behind in innovation relative to the rest of the Brazil. The regions that gained from technological innovation as it occurred are the Center-west, and, surprisingly, the Northeast.

In the Center-West the growth in agricultural income does not come as a surprise given the large productivity improvements in annuals that has been reported in the region. Of the two types of innovation presented in Table 5.5, the more capital-intensive innovation is better suited to transformation of the annual and livestock activities in the Center-west, leading to the highest regional increase in agricultural income (13.6 percent). Technological change in the Center-west favors mainly large farms, and in this respect the growth that has occurred appears to favor land concentration in a region of Brazil with already the most concentrated land ownership in the country.

What does come somewhat as a surprise is that the improvement in income in the Center-west is not larger given the magnitude in productivity improvements in annuals and livestock. The reason for the limited increase in income is linked to the improvements that occurred contemporaneously in the South/Southeast region where the agricultural returns decrease. In these two regions the deteriorating terms of trade for both small and large farms, resulting from the contemporaneous increase in productivity in annuals, limits the income generation impact of the productivity improvements. Given that the products are highly substitutable, the market-clearing price decreases markedly as technological improvement occurs. In the South/Southeast the overall impact is to reduce small farm incomes by 5–6 percent, and large farm incomes by 15–17 percent.

The increase in income in the Northeast, which ranges from 2.3 to 3.7 percent depending on the type of technological

Table 5.6 Change in per capita agricultural income associated with non-Amazon technological change: Scenario replicating innovation for 1985–1995

	Historical		Historical without Livestock technological change		Historical without innovation in the Northeast		Historical without small farm innovation	
	Balanced	Labor & cap	Balanced	Labor & cap	Balanced	Labor & cap	Balanced	Labor & cap
Amazon								
Small Farm	–18.3	–21.5	–2.9	–1.2	–17.8	–19.6	–11.7	–18.0
Large Farm	–27.5	–33.9	6.2	7.8	–25.8	–31.3	–23.4	–30.3
Forest Activities	–23.6	–26.5	20.0	27.6	–20.7	–24.3	–15.6	–21.7
Regional total	–21.9	–26.1	3.0	5.5	–20.7	–23.9	–15.9	–22.4
Northeast								
Small Farm	10.1	6.7	11.5	12.0	–14.5	–16.8	–16.9	–17.1
Large Farm	–7.4	–5.7	8.6	11.9	–22.3	–24.9	17.8	13.7
Forest Activities	0.5	1.4	–2.2	–2.3	4.3	5.8	1.2	2.4
Regional total	3.7	2.3	10.0	11.4	–16.5	–18.8	–4.3	–5.8
Centerwest								
Small Farm	–6.2	5.7	4.6	10.0	–12.0	1.8	–32.6	–16.5
Large Farm	5.1	15.8	1.3	4.4	4.6	17.8	21.6	33.2
Forest Activities	0.5	0.5	–2.0	–2.0	–0.3	–0.8	0.1	0.1
Regional total	2.9	13.6	1.9	5.4	1.4	14.4	10.9	23.1
South/SE								
Small Farm	–5.4	–6.4	0.2	1.3	–0.5	–1.7	–16.9	–15.9
Large Farm	–14.6	–16.7	–3.7	–2.8	–11.7	–13.5	0.0	–6.6
Forest Activities	–0.7	1.4	–3.3	–2.8	–1.4	0.3	–1.1	1.0
Regional total	–8.8	–10.1	–1.5	–0.5	–4.9	–6.2	–9.6	–11.5
National Total	–6.6	–6.8	1.3	2.7	–7.6	–7.6	–6.9	–7.6

change (balanced or capital and skilled labor efficiency improvements), is due to the fact that the innovation that occurred (in perennials and livestock production) is not in competition with the innovation in annuals that occurs in the other two regions. Independently of whether the innovation is balanced in factors or labor&capital improving, small farms appear to perform better even though small and large farms in the Northeast experienced similar rates of technological change. This disparity is due to the assumption we make that capital markets are segmented: small farms and large farms compete for different pools of capital. In the historical scenario, it appears that small farms are able to attract capital from other regions, whereas large farms are not competitive when compared to other large farms (that are also innovating). The uneven income distribution, which is a hallmark of the Northeast, improves under the historical scenario.

Given that livestock innovation outside the Amazon played such a prominent role in limiting deforestation rates, it is worthwhile to analyze what impact livestock innovation had on income generation in agricultural areas. By looking at the third and fourth columns in Table 5.6, one can see that if innovation in livestock had not been introduced, compared to the historical scenario all regions except the Center-west would have been better off in terms of agricultural income, and the outcome would have been more equitable in the Center-west and South/Southeast. If no livestock innovation were to occur in Brazil, the Northeast, which had smaller but more diversified productivity gains than other regions, would experience a considerable increase in incomes for both small and large farms. The income growth in the Northeast would be driven in this case by a shift of resources towards the production of perennials. The large differences in deforestation and income between the scenarios with and without livestock innovation indicate that policymakers are faced with trade-offs to the extent that the desirable environmental outcome associated with technological improvement in livestock outside the Amazon is attained at a cost in both income generation and income distribution objectives.

The type of development that occurred in the Northeast, on the other hand, appears to have been beneficial for both the environment and for income generation. It would lead to less deforestation and greater incomes in Brazil as a whole (besides the obvious increase in income in the Northeast).This can be deduced from columns 5 and 6 in Table 5.6, which show the change in value added that would have occurred if innovation in the Northeast had not taken place: a 17–19 percent decrease in Northeast agricultural income and a 7.6 percent decrease in Brazilian agricultural income relative to the 1995 observed values. Even if no livestock innovation were to occur in Brazil it appears the impact in environmental terms of innovation in the Northeast would be inconsequential. From the results presented, one may state that innovation in the Northeast could potentially be a win-win scenario, with the extent of the environmental improvement depending on the degree of innovation in livestock that may occur.

Small-farm technological change outside the Amazon played an important role in achieving both income and equity objectives. The last two columns of Table 5.6 illustrate what would have happened to agricultural income if large farms alone had innovated, leaving small farms behind: there would have been a predictable decrease in small-farm income but also a decrease in national agricultural income, relative to the historical scenario. Here too policymakers may have an appealing option to the extent that stimulating technological innovation among small farms would generate income, improve income distribution, and reduce deforestation.

To conclude this section, it is important to highlight the importance of livestock technological innovation in determining the outcome both in terms of impacts on incomes and on deforestation. Technological improvements in the Northeast and among small farms are potential win-win scenarios; however, the deforestation reductions are contingent on innovation in livestock production occurring. Without innovation in livestock outside the Amazon, deforestation would increase even with improvements in other technologies in the Northeast and among small farms. At the same time, innovation in livestock has negative impacts on agricultural income. It therefore appears that this is an unavoidable tradeoff, albeit one that may not be under the policymakers' direct control.

Decomposing Productivity Improvements Outside the Amazon (by activity and by region)

The previous section highlights how agricultural technological improvement outside the Amazon was greatly diversified both regionally and in terms of the activities used to carry out the improvements. Different improvements had conflicting effects on the deforestation rate and on income. In particular, livestock improvements were crucial in avoiding the increase in deforestation rates that accompanied technological improvement in annuals in Center-West and South/Southeast. However, this containment of deforestation rates came at a cost in equity and income-generation objectives.

This illustrates how decomposing the impact of technological change by region and by activity may help in understanding future effects of agricultural technological change on deforestation and agricultural income.

The simulations performed in this section consider technological change separately by type of activity and region. Figure 5.13 presents the impact on deforestation rates of each type of technological change (annuals, perennials, and livestock) by the region where innovation takes place (Center-West, South/Southeast, Northeast). Innovation occurring simultaneously in all regions is also considered, along with innovation occurring at the same pace in all activities. The results show quite clearly that where technological change occurs outside the Amazon has less weight in determining the rate of deforestation than the type of activity in which it occurs. So, for example, improvements in livestock technology outside the Amazon consistently decrease deforestation no matter where they occur, whereas innovations in annuals or perennials outside the Amazon appear to have the opposite effect on deforestation, leading to an increase in deforestation rates.

The reason behind the importance of the type of innovating activity, rather than where the innovation is occurring, is that the mechanism transmitting the productivity shock lies mainly in the terms-of-trade effects that arise for Amazonian products as a consequence of the resource shifts induced by technological change outside of the Amazon. For example, if the South/Southeast innovates in the production of annuals, resources will be attracted away from South/Southeast livestock and perennials, causing terms of trade for these activities to improve in the other regions. Since livestock in the Amazon makes such extensive use of land, the change in the terms of trade will cause demand for pastureland to increase, and therefore deforestation will increase. For the same reason, but in reverse, if technological change occurs for livestock production outside the Amazon, the terms of trade for livestock producers in the Amazon will deteriorate and demand for pastureland in the Amazon will decrease, ultimately causing deforestation rates to decrease.

Given the conflicting effects of improvements in livestock, as opposed to those in annuals or perennials, one would expect that innovation occurring at the same pace in all three activities would lead to an outcome for the deforestation rate that is in between the decrease associated with livestock improvement and the increase encountered with annuals improvement. However, this is not the case: when all three types of technological change occur simultaneously, the effect is to consistently decrease deforestation even more than the livestock improvement scenario. This occurs because (1) the terms of trade for *all* Amazonian agricultural products deteriorate due to increased production in the region where the innovation takes place, and (2) the region where innovation occurs now attracts factors interregionally, rather than redistributing them internally. Therefore, the reason that this option is effective in slowing deforestation is that no single factor or activity is pushed into the frontier.

To conclude, it is clear from the results that "balanced" technological change outside the Amazon, where all factors become more productive in all agricultural sectors, is the option that most reduces deforestation rates. At the extreme, if balanced change were to occur at the same pace in all regions outside the Amazon, it would translate into a substantial decrease in deforestation rates, given that there is a 2 percent reduction in the deforestation rate for every 1 percent improvement in TFP (Figure 5.13). However, one must consider the effects on income of such widespread technological change in agriculture, and this will occur mainly through terms of trade for agricultural goods produced in the different regions.

Figure 5.13 Change in deforestation rates for non-Amazon technological change: Decomposing the impact of innovation among activities

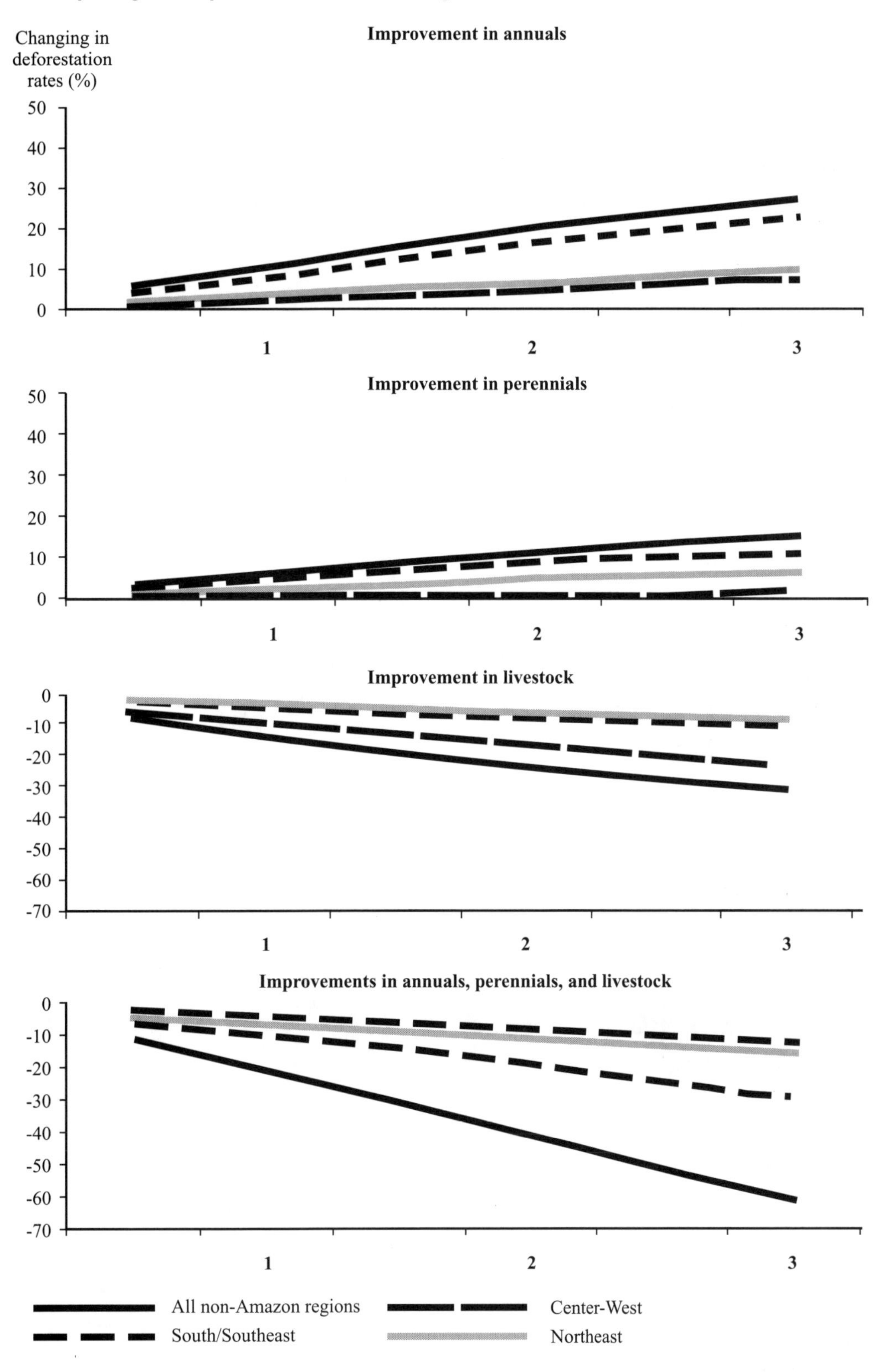

Note: The x-axis is the TFP index as defined for Amazon technological change scenarios.

Figure 5.14 Change in per capita regional agricultural income: Decomposing the impact of innovation by type of activity and region where it occurs (%)

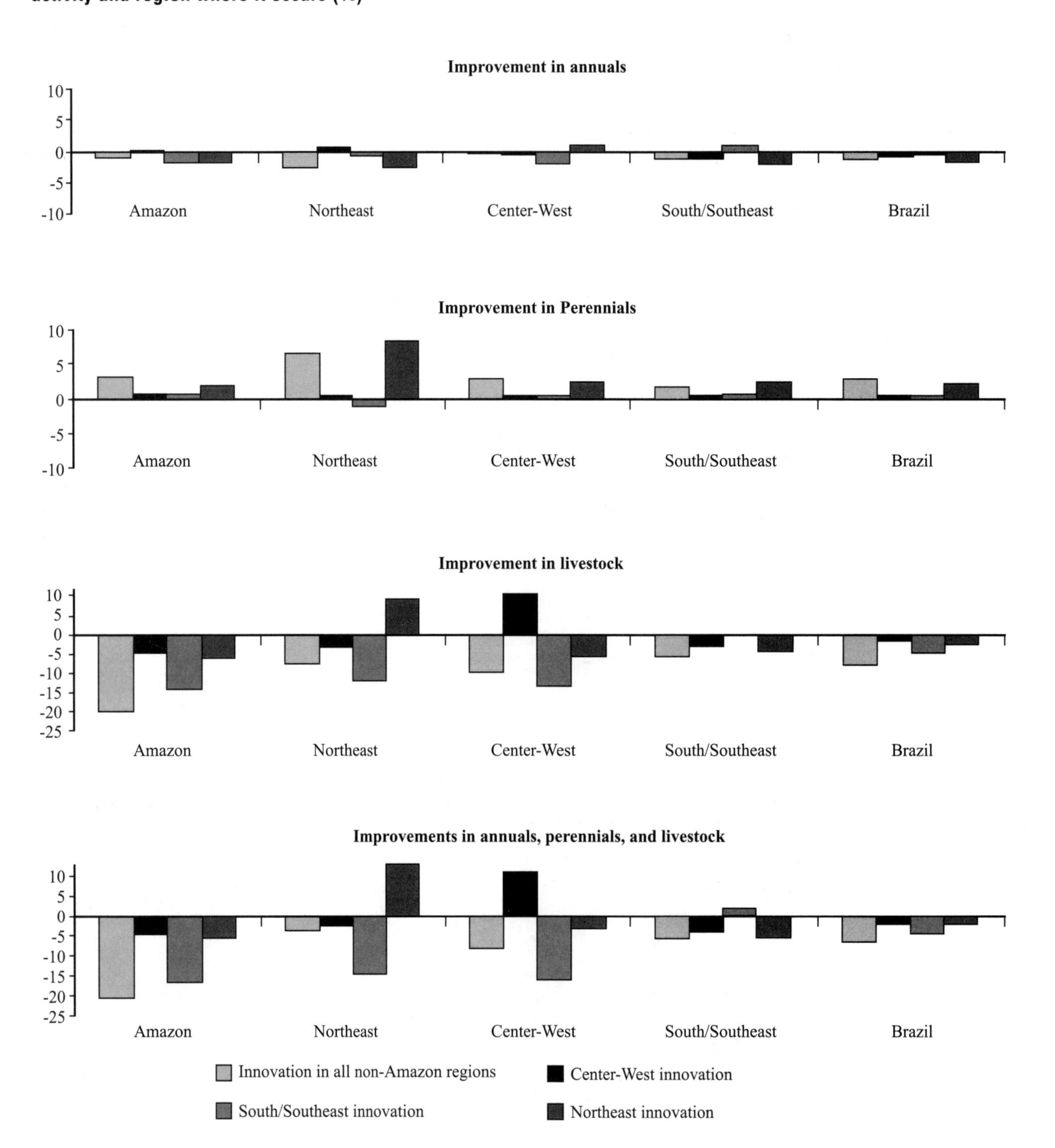

On the income side, innovation in perennials consistently has the best income potential, generating a gain in national agricultural income independent of where it occurs (Figure 5.14). It has the most positive effect, however, when it occurs either in the Northeast or in all three regions at the same time. This is because perennials production in Brazil is quite diversified and produces regionally specific commodities. These commodities, by reason of their regional nature, are less substitutable than annual or livestock commodities that are produced in all regions; therefore, the increased production reverberates less on the terms of trade. Furthermore, the internal reallocation of resources to produce more perennials where the innovation occurs improves the terms of trade for other goods both in the innovating region and elsewhere. This causes regions that do not innovate to increase their agricultural income.

For improvement in annuals, the effect on agricultural income is quite different (Figure 5.14a). Because annuals are highly substitutable, the terms of trade of commodities such as rice, soy, and manioc, which are produced throughout the country, are negatively affected by the goods being produced with improved technologies. This implies that the efficiency gains are counterbalanced by the lower prices paid to producers, leading to very small gains in income in the innovating regions and usually losses in the other regions (which also produce annuals). The internal reallocation of resources mentioned in the perennials case, which improves the terms of trade for the other goods, is not enough to counterbalance the negative impact on income of the price of annuals.

The results indicate that improvement in livestock technologies has a negative effect on agricultural income at the national level, even more so than improvements in annuals. Regionally, if the innovation occurs in the Northeast or the Center-West, it has a considerable (10 percent) positive impact on agricultural income in those regions; otherwise the impact is negative (Figure 5.14c). This supports the general conclusion of the previous section that, although livestock technology improvement outside the Amazon reduces deforestation, it also comes at a cost in terms of forgone agricultural income. Whereas the previous section looked at the impact of removing livestock from the historical (or business-as-usual) technological change, here the scenario analyzes the impact of isolated changes in livestock technology. The greater detail, in terms of what region innovates livestock technologies, indicates that the statement mentioned in the previous section has to be qualified: regional income may improve in some regions depending on how regionally concentrated the innovation is. What also emerges, as long as livestock technologies improve outside the Amazon for both small and large farms, the income distribution may actually improve (see Appendix G, Table G1).

The improvement scenarios consider in detail the possible changes arising from innovation in a single type of activity, whether annuals, perennials, or livestock technologies, but they do not take into consideration the interaction that occurs when multiple activities are innovating. In Figure 5.14d, the impact of this interaction is accounted for by annuals, perennials, and livestock technologies improving at the same time and by the same amount. In this case, the results are similar to those in the livestock improvement scenario of the preceding paragraph (Figure 5.14c). The only difference is that, when the innovation is regional and in three activities, rather than just one, the region gains an extra advantage over regions that are not innovating. However, even though there are regional income gains for the innovating regions, the overall impact on agricultural income at the national level is negative. Finally, the tradeoff between reducing deforestation and meeting income objectives is an important aspect of any policy decision that may affect technological change outside the Amazon.

Incentive Mechanisms for Natural Resource Management: Bridging the Gap between Private and Social Costs

It is often argued that compensatory mechanisms, either in the form of taxes or transfer payments, are necessary to sustain natural forest management (NFM)[44] and conservation market values because there are at present insufficient incentives for long-term forest uses (Richards 2000). The lack of incentives for NFM enables alternative land uses, including one-off logging and conversion to agricultural land, to seem more attractive. A recent assessment of forestry options in Latin America by Southgate (1998),

Box 5.2 The role of land use in curbing carbon emissions

The potential role of forestry and land use in holding carbon captive and sequestering atmospheric carbon is large. The Second Assessment Report of the Intergovernmental Panel on Climate Change estimates it at about 100 million metric tons, or roughly 30 years of fossil fuel releases at current emission rates. The "tradable" component is obviously much smaller, and, in fact, estimations by Pearce et al. (1998) indicate that the market for developing-country offsets is likely to be in the range of 60–120 million tons of carbon. Based on U.S. government estimates of the price of traded carbon ($14–23 per carbon ton), this suggests an annual value of tropical forestry carbon offsets ranging from $840 million to $2,760 million under the Clean Development Mechanism. Other estimates, including that carried out for the World Bank by Ellerman, Jacoby, and Decaux (1998), are more optimistic, indicating that within 10 years a global market in emissions trading could be worth tens of billions of dollars annually, a "substantial percentage" of which should flow to developing countries (Moura Costa et al.1999).

Carbon "leakage" and project permanence are two common objections to the use of forestry sinks that arise from these unresolved issues. Leakage occurs when carbon is lost elsewhere in the system as an indirect result of a carbon offset project. A typical example is when a conservation project leads to an increase in deforestation outside the protected area. In this case the net mitigation believed realized by the initial protection project is reduced or eliminated.

Permanence relates to the concern that a particular land use or forestry condition may be subject to rapid carbon releases. A new forest that is created to sequester carbon could quickly release carbon, either intentionally, through timber harvesting or land clearing, or inadvertently, as with a fire. This problem can be overcome by a contractual arrangement for the credits to be provided on a short-term basis through adoption of a ton-year accounting system, where credit would be given for the number of tons of carbon held out of the atmosphere each year, rather than on the basis of "permanent" sequestration. This would allow comparisons to be made between forest reserve creation and policies to slow deforestation. Ton-year accounting is also needed to compare the amounts of fossil fuel emissions avoided through silvicultural plantations and other mitigation options in the forest sector (Moura Costa and Wilson 2000). The general concept of the ton-year approach is in the application of a factor to convert the climatic effect of temporal carbon storage to an equivalent amount of avoided emissions. In this context, avoided CO_2 emissions from averted deforestation are assumed to be equivalent to avoided CO_2 emissions from industrial sources. Carbon accounting systems for the two types of avoided emissions should therefore be commensurable.

[44] Guidelines for NFM provided by the Rainforest Alliance state that forest operations must maintain environmental functions, management planning and implementation must incorporate sustained yield concepts, and all activities must have a positive long-term impact on local communities.

including "high-value" nontimber forest products, bioprospecting deals, ecotourism, and so forth, concluded that NFM and conservation are not competitive with other uses even at low discount rates.

High discount rates associated with high risks encourage forest mining as opposed to NFM. In the Brazilian Amazon, forest management was found to be unattractive at any discount rate above 1 percent (Verissimo et al.1992). Another problem is the slow growth in forest product prices. Southgate (1998) points out that timber prices in the Amazon are depressed because the supply of timber, much of it illegal and from unmanaged natural forests, is still so abundant.

Two categories of corrective actions for deforestation are considered here. In the first, fiscal mechanisms create disincentives to deforest, while in the second, payments compensate producers for the forgone profits associated with reduced emissions. Fiscal instruments, such as a tax per ton of carbon emissions or a tax on logging, aim to correct market incentives, so that the externalities involved are taken into account. A transfer payments approach involves the transfer of costs between different stakeholders: it mainly compensates landowners, whether public or private entities, for conserving forest. These payments can be made either from domestic sources as payment for national public goods or as international transfer payments for global positive externalities of forest conservation. For the purposes of this report, market or trade-based solutions are included in this category.

Taxation for Conservation Goals

Taxes reflecting the nonmarket benefits and costs stemming from different types of land cover encourage users to move toward more sustainable resource management. Several possibilities are available to legislators:

(1) land use taxes incorporating the benefits or costs from different land use systems can be adopted; a carbon tax or subsidy based on the carbon emitted or sequestered by each land use can be considered an example of this approach, as can a logging tax;

(2) taxing land as an asset can discourage deforestation linked to land speculation by raising the cost of holding land as a hedge against inflation or as a source of capital gains (Kaimowitz, Byron, and Sunderlin 1998).

The tendency of land use taxes, where they exist, has unfortunately been in the opposite direction; in the past Brazil's rural land tax, which was designed to stimulate rural productivity, was found by Almeida and Uhl (1995) to be light on ranching, thus encouraging deforestation. A law was introduced in 1996 that corrected these distortions by exempting untouched forest areas from entering the taxable base. However, there are still no explicit incentives for loggers, ranchers, and farmers to make productive use of forested land using sustainable management techniques, rather than just leaving it unproductive. Few countries have tried land or capital gains taxes due to the large amounts of information required, the high potential for evasion, and the likely political opposition (Kaimowitz, Byron, and Sunderlin 1998).[45]

Transfer Payments to Promote Conservation

What Conservation Options are Available?

Payment schemes for producers adopting sustainable forest management techniques aim to obviate the opportunity cost of

[45]In the near future, global positioning systems and geographic information systems should allow local governments to monitor land use change in a cost-effective manner on a property-by-property basis.

forgoing more profitable activities, which would otherwise lead to deforestation. As mentioned in the introductory section of this chapter, the financial backing may be domestic or international.

An example of a domestic approach is the Brazilian ecological value-added tax (VAT), introduced by four states since 1992, following state legislation to reallocate the VAT according to environmental criteria. The ecological VAT is distributed to municipalities according to the extent to which they favor land uses that encourage conservation and water protection (Seroa de Motta 1997). For example, in the state of Rondônia, 5 percent of the VAT has been distributed to municipalities that protect forests. The mechanism explicitly recognizes the need to compensate municipalities for forgone income, and payments are linked to well-publicized environmental performance indicators. Large increases in municipality participation in the program, resulting in greater funds, have been reported by Seroa da Motta (1997).

There is a wide array of possible international transfer payment arrangements. These can be subdivided into market and nonmarket transfers of financial resources. The latter transfer resources from consumer nations to conserving nations, recognizing that forests are a global public good. The former is driven by profitability once the public good component of forests has been internalized in financial markets.

Among the available nonmarket mechanisms are (1) the Global Environmental Fund, which is responsible for the financial implementation of the International Convention on Climate Change and Biological Diversity. It has provided Brazil with approximately US$30 million, and another Fund project, to begin in 2002, is expected to contribute another US$30 million; (2) the World Bank, which set up a $150 million Prototype Carbon Fund, is buying carbon emissions reductions in the amount of US$13.6 million; and (3) the Pilot Program to Conserve the Brazilian Rain Forests, funded by the G-7 countries, the European Union, the Netherlands, and Brazil itself, has provided about US$340 million for the period 1992–2002, out of concern for the deforestation of Brazil's humid rainforests.

Among the market-based approaches, biodiversity prospecting and carbon trading are the two main options. Biodiveristy prospecting has generated considerable hope in recent years, but it appears that initial estimates of the commercial value of conserving forests for pharmaceutical purposes were overoptimistic. The expected value per hectare of bioprospecting is reported to be an order of magnitude smaller than the opportunity cost of holding land (Aylward 1993; Simpson, Sedjo, and Reid 1996; Southgate 1998). Overall, the approach does not appear to present substantial opportunities for conservation in the Brazilian Amazon because of the uncertain financial returns, the large area involved, and because funds would probably be directed to the Atlantic coastal forest of Brazil, which is at higher risk.

There has been increasing optimism surrounding carbon trading based on the accelerating political process of establishing binding carbon emission limits. In December 1997 a number of countries agreed to reduce their carbon emissions as a first step toward halting global climate change. The Kyoto Protocol, once ratified, will be the first legally binding international agreement that sets targets for cutting greenhouse gas emissions. Article 12 of the Kyoto Protocol gave a major boost to carbon trading by establishing the Clean Development Mechanism (CDM). The CDM was included in the Protocol as a proposal from the government of Brazil to create a means whereby countries not accepting binding emissions limits could cooperate on a project-specific basis with countries that had agreed to limitations (Annex I countries) in reducing emissions. The CDM calls for real, additional, and cost-effective reductions of net carbon emissions. The forest sector in Brazil offers considerable scope for activities within the CDM, including opportunities for private sector investors. However, a number of institutional and policy mechanisms must be established by the government and international agencies to ensure that these activities meet the objectives of the CDM. In particular, there is continuing dissent among the parties concerning carbon sinks (areas such as forests that store or hold carbon) in general and whether forestry will be included in the CDM, given that it is not specifically mentioned in Article 12.[46]

Opportunities for Slowing Deforestation Under the Kyoto Protocol

> Any reduction in the rate of deforestation has the benefit of avoiding a significant source of carbon emissions [...] Limiting deforestation forgoes the opportunity to utilize the land for other purposes, such as agriculture or other developed uses, therefore would potentially be subject to the same opportunity costs that might arise with afforestation and reforestation.—*Land use, land-use change, and forestry: A special report of the IPCC* (Watson 2000)

The discussion that follows explains the place of Brazil in combating global warming, outlining opportunities presented by the country's forest sector, and the obstacles that must be overcome to turn these into global warming response options. Unsettled issues in assigning credit for carbon include deciding whether carbon is counted on the basis of permanent sequestration versus carbon ton-years, agreeing on methods for crediting forest reserve established, applying discounting or other time-preference weighting systems to carbon, and deciding whether credit will be based on avoided emissions or on stock maintenance (see Box 5.2).

Two approaches are frequently mentioned in proposals to use tropical forest maintenance as a carbon offset. One is to set up specific reserves, funding the establishment, demarcation, and guarding of these units. Monitoring, in this case, consists of the relatively straightforward process of confirming that the forest stands in question continue to exist. In the Brazilian Amazon, where large expanses of forest do still exist, the reserve approach has the logical weakness of being completely open to "leakage": that is, with the implantation of a forest reserve, the people who would have deforested the area established as a reserve will probably clear the same amount of forest somewhere else in the region.

The second approach is through policy changes aimed at reducing the rate of clearing in the Amazon region as a whole (not limited to specific reserves or areas of forest). This second approach has the great advantage of addressing more fundamental aspects of the tropical deforestation problem, but it has the disadvantages of not assuring the permanence of forest and of not resulting in a visible product that can be convincingly credited to existence of the project. In order for credit to be assigned to policy change projects, functional models of the deforestation process must be developed that are capable of producing scenarios with and without different policy changes.

The manner in which carbon credits are calculated can determine whether policy change mitigation options are subject to leakage or negation of the carbon benefits by events outside a given project area set in motion by the mitigation activity. Because the policy change approach focuses on national-level totals (whether these totals be

[46]Article 3.3, of the Kyoto Protocol, however, does specifically consider afforestation, reforestation, and deforestation as activities that affect the level of carbon in the atmosphere and that can be used as tools to reduce a country's level of carbon dioxide releases.

of flows or of stocks), no leakage can occur through changes in the spatial distribution of deforestation activity within the country, as by movement of potential deforestation from a reserve to another forested area. Displacement of deforestation in time, however, can result in leakage if the accounting procedure requires permanent sequestration in either specific areas of forest or in the forest sector of a whole country. In essence, credit for efforts to combat deforestation will require (1) acceptance of contributions to larger programs, rather than restricting recognition to free-standing projects, and (2) adoption of a ton-year accounting system for carbon so that contractual arrangements for credits could be provided on a short-term basis.

Evaluating the Potential Impact of Incentive Mechanisms in Brazil

The previous sections looked at the available options for reducing the gap between private and social benefits of maintaining forest stocks. This section compares the impact of the different policy options on both deforestation rates and income distribution. The options are

(1) fiscal instruments to account for the externalities involved:

- a deforestation tax at rates of R$30, R$40, and R$50 per hectare is imposed (this is equivalent to a carbon tax rate of R$0.14, R$0.20, and R$0.25 per carbon ton);
- a tax on logging output from the Amazon region is introduced in 5 percent increments (5, 10, and 15 percent);

(2) transfer payments involving the transfer of costs between different stakeholders for conserving forest: in the simulation this is represented as a government subsidy to the extractive component of forest-related activities. These payments could, in fact, be made either from domestic sources (as payment for national public goods) or as international transfer payments for global positive externalities of forest conservation. The forest subsidy scenarios are obtained for 10 percent increments in the subsidy rate (10, 20, and 30 percent subsidies on value of extractive activities).

For fiscal instruments to be set correctly to be effective and equitable, research is needed on the difference between the private and social costs of the different winners and losers, and on the marginal costs of the resource users (Markandya 1997). This section provides some insight into the effectiveness of these different measures. When interpreting the results, one should remember that the measures under investigation also demand considerable administrative capacity including monitoring, enforcement and collection, and the need for wide public consultation prior to their introduction, which are not part of the analysis here.

From the results presented in Figure 5.15, it appears that a tax on logging activities (for tax rates up to 15 percent) does not lead to a decrease in the deforestation rates, even though the model takes into consideration the link between logging and deforestation. On the contrary, the reduction in logging causes resources to be shifted toward deforestation for agricultural purposes; however, this effect is only minimal due to the complementary nature of deforestation and logging. As one would expect, this policy has a considerable negative impact on the logging industry and raises substantial revenue. A 15 percent tax rate on logging output would reduce output by R$80 million, while raising fiscal revenue by R$53 million (Figures 5.15, and 5.16). It is interesting to observe that the negative impact at the local level is somewhat compensated for by improved terms of trade for logging in the other regions of Brazil (Table 5.7). Overall, if the objective is to reduce deforestation rates in an equitable manner, introducing a logging tax will not accomplish the goal.

The deforestation tax scenario appears more promising than a logging tax, as is to be expected because it is a more targeted approach, linking directly to the externalities arising from deforestation. If the tax on deforestation activities was set at R$50 per hectare, the yearly deforestation rate would be reduced about 9,000 square kilometers, with logging being only minimally affected. Extractive activities would stand to gain from this tax and would expand output by about R$60 million—a 25 percent increase (Figure 5.16). The revenue-raising potential, assuming such a tax could be collected effectively, is similar to that of the logging tax presented previously. At the highest rate considered (R$50 per hectare), the revenue generated by this fiscal measure is R$60 million (Figure 5.17). The results of this scenario are different from the logging tax scenario in several respects: a substantial decrease in the deforestation rate occurs with the deforestation/carbon tax, but the negative welfare effects of introducing a deforestation tax are considerably higher (see the change in regional agricultural value added in Table 5.7).

As reported in Richards (1999), tax rates correcting for negative externalities have often been set too low, possibly as a result of political resistance and lack of research, but also of confusion between the incentive and revenue objectives. An incompatibility of these objectives is pointed out by Karsenty (2000): in order to achieve an environmental impact by correcting economic behavior, the charge needs to be set at a high enough level and it must be narrowly targeted, whereas for revenue generation a lower charge and a broad tax base are better. An example of this incompatibility is the deforestation "contribution" levied by the Brazilian Federal Environmental Agency (IBAMA) on small operations consuming less than 12,000 cubic meters of forest raw material (as opposed to carrying out the mandatory forestry reposition equivalent to the consumption level).[47] Such a policy aims only to achieve revenue generation objectives, with part of the revenue diverted to reforestation activities; however, the required payments (US$4.00 per cubic meter) were not high enough to modify deforesters' behavior (Seroa da Motta 1997). Stone (1998) reports prices ranging from US$24 to $82 per cubic meter: therefore, the tax is equivalent to a 5–15 percent tax rate depending on wood quality. The taxed activity-effectively logging (since what is being taxed is volume of wood abstracted)—has only an indirect link with deforestation for agricultural purposes, making it an ineffective policy for reducing deforestation rates. It is much more effective to target hectares that are being deforested for agricultural purposes, because such a low value-added deforestation activity can be deterred even by politically feasible tax rates, in this case even without forgoing any revenue-raising potential. (The monitoring and collection of the tax may be difficult to accomplish, however.) To understand the efficiency of a deforestation tax in the context of carbon emissions reduction, it suffices to say that a tax of R$50 per hectare is equivalent to a carbon tax of R$0.25 per carbon ton, which is much smaller than any tax rate being proposed in

[47]The Brazilian Forestry Code states that those exploiting or utilizing forestry raw materials are obliged to plant appropriate species, at some location, equivalent to the amount the exploiter consumed. This requirement covers logging as well as consumption of charcoal and firewood of unknown origin. Since 1978, however, a federal norm allows those consuming less than 12,000 cubic meters of forest raw material per year the option of paying a deforestation contribution, instead of investing in reforestation.

Figure 5.15 Impact of tax and subsidy scenarios on deforestation, logging, and extractive activities

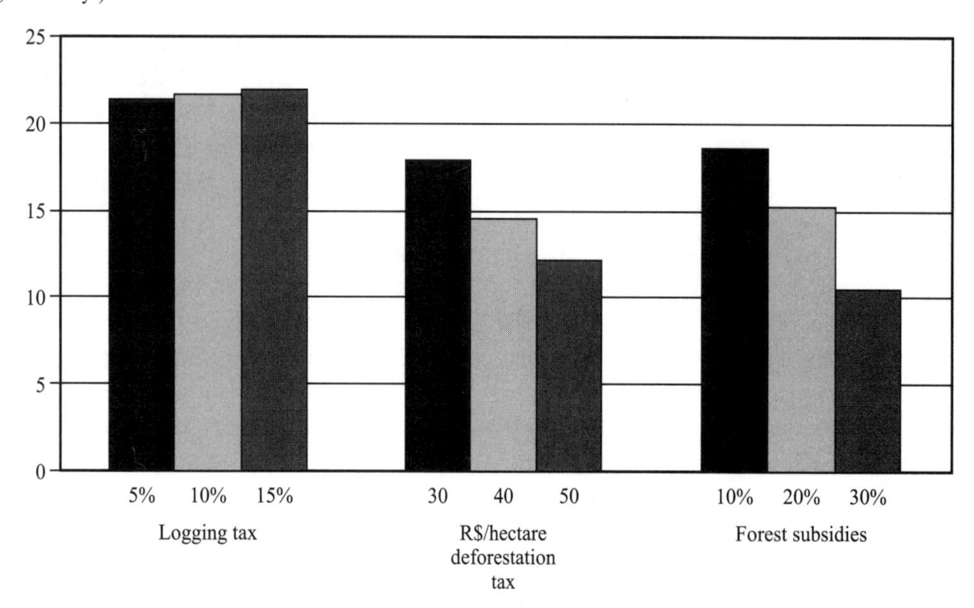

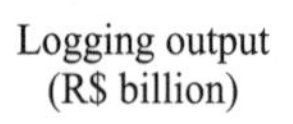

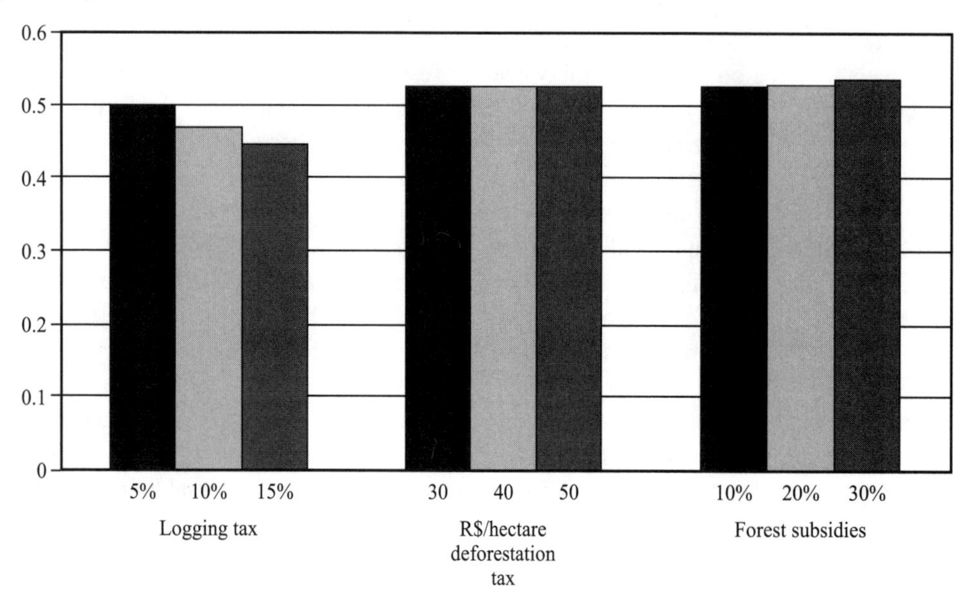

Figure 5.16 Impact of tax and subsidy scenarios on extractive activities and government revenue (Rs billion)

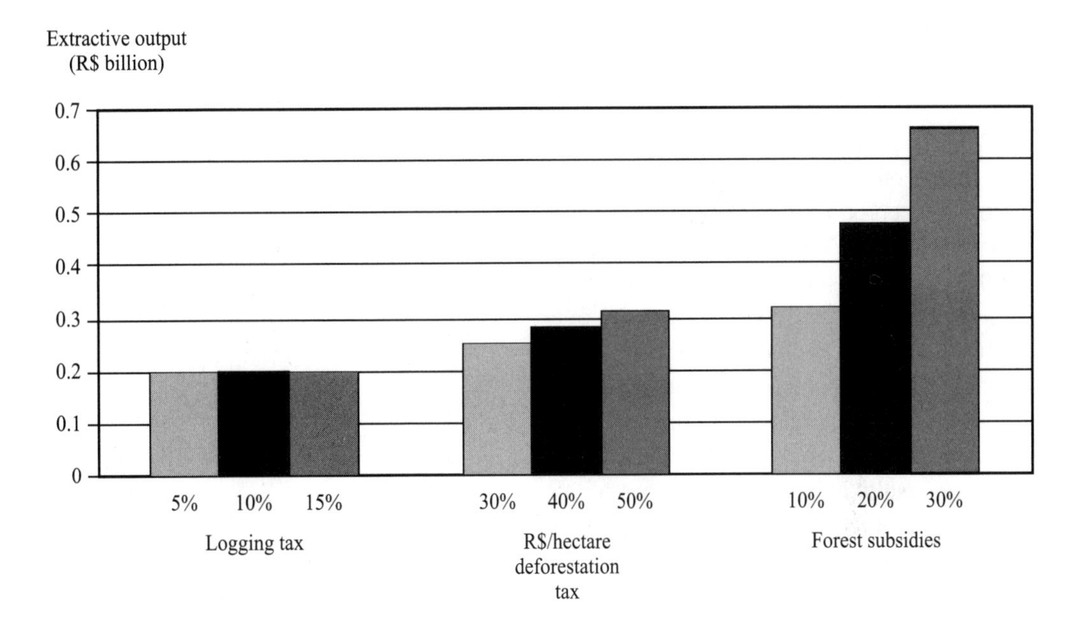

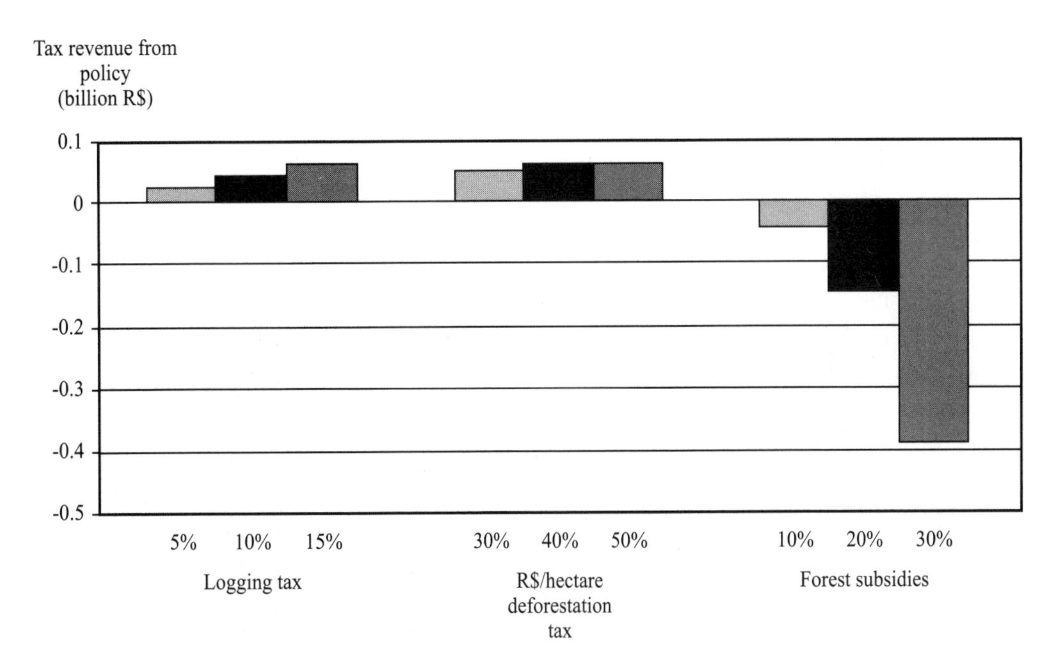

developed countries to curb emissions. It therefore appears that, if such Amazon-wide reduction in deforestation rates were to be allowed under the Kyoto Protocol, there would be interesting options for CDM trades.

The third and final scenario aims to support forest conservation by subsidizing nontimber forest product extraction in the Amazon region. Vosti, Witcover, and Carpentier (2002) report that subsidizing forest conservation on small farms with payments of R$100 per hectare would reduce the deforestation rate substantially on these farms (by 36 percent on average). In the broader Amazon context of the model presented here, the results indicate that a higher subsidy would be required to obtain comparable reductions in deforestation rates. The results represent a 10, 20, and 30 percent subsidy for nontimber forest products. These are equivalent to a R$150, R$240, and R$360 per hectare subsidy per hectare of forest spared. In particular, it would take a R$240 per hectare subsidy to obtain a 30 percent reduction in deforestation (a R$150 per hectare subsidy would lead to only a 12 percent reduction). The discrepancy between Vosti, Witcover, and Carpentier (2002) and the numbers reported here may be imputed to several factors: (1) the deforestation carried out on larger farms is also considered here; (2) the speculative component of deforestation, included in the model, has an impact on the effectiveness of the policy (only tenured farmers would be entitled to subsidies); and (3) the simplistic approach to time in the model presented here may account for part of the difference (the deforestation rate in Vosti, Witcover, and Carpentier 2002 had to be averaged over time to obtain a rate reduction comparable to the one in the results presented here). These differences notwithstanding, conservation subsidies would be a popular policy and could lead to substantial reduction in deforestation rates.

Table 5.7 Percent change in income to different types of producers based on the tax and subsidy scenarios

Region/activity	Logging tax (%)	Deforestation tax (R$ 70/hectares)	Forest subsidy (30%)
Amazon			
Small farm	–0.30	–0.70	0.08
Large farm	1.69	3.20	–4.88
Forest activity	–9.87	–19.94	42.87
Regional	–0.96	–2.06	4.27
Northeast			
Small farm	0.13	–0.05	0.22
Large farm	0.12	–0.13	0.34
Forest activity	4.92	2.12	6.50
Regional	0.30	0.00	0.50
Center-West			
Small farm	0.36	0.18	0.47
Large farm	0.30	–0.18	0.42
Forest activity	5.20	1.36	4.41
Regional	0.42	–0.08	0.51
South/Southeast			
Small farm	–0.10	–0.07	0.28
Large farm	0.14	–0.18	0.37
Forest activity	11.17	1.85	6.17
Regional	0.48	–0.03	0.58

From a welfare standpoint, all regions stand to gain from the subsidies: the Amazon as the direct beneficiary but the other regions indirectly as a result of the decrease in the volume of wood coming out of the Amazon (Table 5.7). Benefits accruing to the Amazon with a 30 percent subsidy, after accounting for price changes caused by the subsidy, are in the range of R$213 million, while indirect benefits to the other regions from increased output in logging activities are R$268 million. Hence the subsidy expenditure, which appears high at R$388 million, is more than compensated for by the R$481 million in market benefits to be accrued nationwide. However, the opportunity cost of raising the money to pay for the subsidies is not considered here, which would almost certainly be considerable. This is not a crucial point, however, since foreign funding, linked to deforestation's global externality component, may be available to pay these subsidies.

From the previous section on transfer payment options open to Brazil, it appears that over the last decade a substantial amount of funding has been made available to reduce greenhouse gas emissions. However, in terms of annual payments, the amount translates into less than US$50 million, which would not substantially reduce deforestation rates, according to the results of this study. As in the case of the deforestation tax, if an Amazon-wide reduction in deforestation rates were allowed under the Kyoto Protocol, the introduction of the conservation subsidy would provide interesting options for CDM trades. The subsidy is equivalent to a payment of R$1.21 per carbon ton, which is again much smaller than the marginal cost of reducing emissions in developed countries.

CHAPTER 6

Policy Conclusions

The policy implications of the research presented in this report are far-reaching. The first set of simulations, devaluing the Brazilian real (R$) from 10 to 40 percent, finds that logging in the Amazon would rise with a fall in the currency, because the increase in wood exports would more than compensate for the decrease in demand for wood in the domestic market. The only way the government could avoid such a crisis in the wake of a devaluation would be to address this issue directly by imposing taxes on wood exports or extending the ban on certain types of hardwoods to more species.

In reaction to a currency crisis, however, deforestation to clear agricultural land would be very sensitive to the aggregate behavior of the economy. The balanced-adjustment scenario, with a reduction in equal shares of private consumption, government demand, and investment would lead to a reduction in deforestation in the short run and a small positive increase in the long run. The capital-flight scenario, where government expenditures and households savings rates are left unchanged (which would decrease investment drastically) would lead to a small increase in deforestation in the short run and a substantial increase in the long run. Under this scenario, a 40 percent devaluation would cause, in the long run, a 20 percent increase in deforestation, meaning an increase of approximately 4,000 square kilometers in the annual deforestation rate.

In the short run, in both the balanced-adjustment and the capital-flight scenarios, rural households would stand to gain and urban households to lose from a currency devaluation. The short-run result is intuitive to the extent that agriculture can shift production between exportable and domestically consumed goods, while some nonagricultural sectors cannot. In the long run, it is still true that incomes would rise in rural areas and fall in urban areas; however, rural per capita income appears to decrease, since those who are worst off in the urban sector of the economy would be the ones to migrate. From a policy standpoint, the balanced-adjustment scenario, which contains final demand, would be less likely to redirect resources from the urban to the rural side of the economy. This is a desirable effect because it shows that the productive capacity present in the urban side of the economy would not be dismantled: instead the savings rate would adjust to the exit of foreign capital. In aggregate growth terms, the crisis affects GDP substantially no matter what course of action is taken. The results indicate that a devaluation of 30 percent would lead to a 3.5–4.0 percent reduction in GDP in both the short and the long run.

The remaining sets of simulations are intended to investigate possible changes in the current structure of the economy that would be pro-active rather than reactive to a crisis. These changes could be directed either toward developing the economy or reducing deforestation, or ideally both. First, the policy relevance of a reduction in transportation costs for agricultural goods produced in the Amazon stems from the changes presently occurring in the region's infrastructure. In all cases, a reduction in transportation costs increases deforestation rates. A 20 percent reduction in transportation costs for all agricultural products from the Amazon increases deforestation by approximately 15 percent in the short run and by 40 percent in the long run. To recast these numbers in terms of hectares deforested per year, a reduction in transportation costs would lead, in the long run, to an increase of about 8,000 square kilometers deforested annually. The return to arable land would increase, thereby increasing the incentive to deforest. A 20 percent reduction in transportation costs would lead to a 24 percent increase in the return to arable land in the short run and a 92 percent increase in the long run.

The increase in profitability of Amazonian agriculture would lead, in the long run (with mobile agricultural labor and capital), to a 24 percent increase in production by smallholders and a 9 percent increase in production by large farms in the Amazon. Welfare effects at the national level, however, are quite limited (nationally rural households would gain 0.6–0.9 percent in real income). This is because the increase in Amazonian production, except for the share that is exported, replaces previous production from other regions; therefore, the positive regional impact on development in the Amazon is offset by the negative impact on other agricultural areas of Brazil.

With the changes that are under way at the national level in macroeconomic policy and transportation costs, it appears that deforestation rates will continue to be high or perhaps even increase in the future. However, since policies at the local level may play an important role in determining deforestation rates, this research looks for local solutions to issues of deforestation and development. The report also analyzes the impact of tenure regimes, agricultural technological change in the Amazon, and possible incentives to reduce deforestation. It concludes that there is no single solution to the issue of deforestation and development in the Amazon, but rather a package of options that could be adopted.

At the Amazon policy level of analysis, regulating tenure regimes and enforcing them is a likely means of reducing deforestation, considering that the Brazilian government has reported extensive fraudulent land claims in the Amazon. This implies that much of the current deforestation is occurring at the hands of untenured deforesters who are cutting down trees in order to occupy the land and thus acquire informal tenure. If land claims were verified by the government and violators evicted, the incentive to deforest to acquire informal property rights would decrease. As the probability of eviction increases, only tenured landholders would have an incentive to deforest, and their motivation would not be speculative but rather the value added that comes from agricultural activities. By removing the speculative incentive to deforest, the deforestation rate could be reduced up to 23 percent.

Whereas tenure regime enforcement aims to remove an institutional distortion, technological change in the Amazon addresses deforestation and development from the standpoint of productivity and the factors of production that are employed in agricultural production. The relative profitability and land intensities of different activities, combined with soil productivity and sustainability limits, are all factors that affect farmers' incomes and determine, in part, the pressures on forests. The adoption of technological change can be influenced and directed by policymakers through the allocation of research and extension funds. To the extent that these simulations represent technically feasible innovations, they are extremely relevant to policy. It is encouraging that the offsetting potential of technological change in terms of deforestation is of the same order of magnitude as the interregional effects on deforestation, if the technologies are carefully chosen. Table 6.1 summarizes the findings on the technological front.

The food security conclusions are based on the author's judgment concerning the production structure after technological change occurs. According to the criteria adopted, if there is specialization in activities with small domestic (Amazonian) markets or prices are volatile, food security would decrease. But if the increase in production is in Amazonian staples, then food security in the region would improve. The best option in terms of food security is innovation in livestock technologies, which increases production of both annuals and livestock. Technological change in annuals is also a good option for food security because the production of staples such as manioc and rice greatly increases, while livestock production is not adversely affected. In this context, perennials are considered risky. The decrease in production of perennials is dramatic only when labor intensification in annuals occurs (which may decrease perennials production by more than 50 percent for high levels of technological adoption). Conversely, technological innovation in perennials leads to specialization in perennials and substantial reductions in the production of annuals and livestock. The perennials land-saving scenario, in the long run, causes a 20–25 percent reduction in annuals and a 30–40 percent reduction in livestock for high levels of technological adoption.

As Table 6.1 shows, the trade-off between forest conservation objectives and agricultural growth is significant in the Amazon. Improvements in livestock technology appear to offer the greatest returns for all agricultural producers in the region, and such improvements should also improve food security. However, improvements in livestock technology would increase deforestation dramatically in the long run.

The alternative would be to pursue improvement in technology for perennials,

Table 6.1 A qualitative comparison of the impacts of technological change in the Amazon

Type of technological change	Land use	Deforestation reduction		Smallholder income (SR)	Large estate income (SR)	Food security (SR and LR)
		SR	LR			
Total factor productivity	Annuals	×××	××××	0	✓	✓✓
	Perennials	✓	×	✓	0	×
	Livestock	×××	××××	✓	✓	✓
Labor productivity	Annuals	✓	××××	✓✓	✓	✓
	Perennials	✓✓	✓✓✓✓	✓✓✓	✓	××
Capital productivity	Annuals	×	××××	0	✓✓	✓
	Perennials	✓✓✓	✓	0	✓✓	×
	Livestock	✓✓✓✓	××××	✓✓✓✓	✓✓✓✓	✓✓✓
Labor and capital productivity	Annuals	✓	×	0	✓✓	✓✓
	Perennials	✓✓✓	✓✓	✓✓	✓	×××
	Livestock	✓✓	××××	✓✓✓✓	✓✓✓✓	✓✓✓✓
Sustainability + labor and capital productivity	Annuals	✓	××	0	✓✓	✓✓
	Livestock	✓✓	××××	✓✓✓✓	✓✓✓✓	✓✓✓✓

Source: ✓ implies a desirable effect (with ✓✓✓✓ as most desirable); × implies an undesirable effect (with ×××× most undesirable); 0 indicates a negligible effect. SR is short run; LR is long run.

because a switch to perennials would intensify labor requirements, which could reduce deforestation rates considerably. The equity effects from improving perennials would be progressive because small farmers would gain the most. If this technology were adopted widely, however, food security would suffer, since fewer farmers would be raising staple food crops. (In addition, farmers would be more exposed to the risks associated with perennial production.) The perennial option has theoretical potential, but the fact that large farms would not be likely to adopt it because their gains would be small, and smallholders would be reluctant to adopt it because they are risk-averse would probably limit the effectiveness of this solution. But even if the option of improving perennials was adopted only in part, it would still help reduce deforestation rates.

Improvement in production of annuals appears to have little potential: in the long run it would reduce deforestation only if land use was greatly intensified, and income effects would be quite small. Before the high level of land intensity required to decrease deforestation rates could be reached, there would almost certainly be, in the early phase of adoption, a period in which these rates would go up substantially.

On a more theoretical note, the results indicate that the type of factor intensification alone does not determine whether deforestation rates will increase or decrease. The factor intensity in the activity being improved is what matters, compared with other activities. Furthermore, the striking difference in deforestation rates between the short and the long run points to the fact that interregional flows of labor and capital play a crucial role in determining the expansion of the agricultural frontier.

Having observed the impact that technological change occurring within the Amazon region can have on deforestation and incomes, it seems appropriate to return to the broader interregional scale to analyze what role technological change outside the Amazon might have played in the past and what significance it may have in the future. The results indicate that, contrary to expectations, the type of agricultural technological change that occurred outside the Amazon during 1985–1995 did not cause an increase in deforestation. In fact, it limited deforestation. This reduction was mainly the result of innovation in livestock technologies outside the Amazon (improvements in planted pasture and in confined animal feeding operations), which occurred alongside improvements in annuals and perennials. The theme underlying this finding is that if production of a good associated with deforestation (such as livestock in the Amazon) is increased outside the Amazon through technological improvements, the terms of trade for the same good produced in the Amazon deteriorates, leading to less deforestation. Along similar lines of reasoning, but in reverse, the increase in productivity of annuals or perennials outside the Amazon causes an increase in deforestation rates.

The impact on per capita income of the type of technological change that occurred outside the Amazon from 1985 to 1995 was that the Center-West and Northeast regions clearly gained in terms of regional income. The income distribution gap apparently decreased in the Northeast and increased in the Center-West as a result of technological change outside the Amazon. The surprising decrease in returns to agriculture in the South/Southeast region was associated with declining terms of trade in annuals.

When considering technological change outside the Amazon occurring separately by region and activity, improvements in livestock or perennials in the Northeast emerge as a win-win outcome in terms of environmental and income objectives. Another option, technological change occurring at the same pace for all agricultural activities outside the Amazon, would cause the largest decrease in the deforestation rate of all, but it comes at the expense of agricultural income. This option is effective in slowing deforestation because no single factor or activity is pushed into the agricultural frontier, but it also has the effect of substantially lowering agricultural prices, with negative consequences for agricultural income.

All the policies discussed up to now apply to the functioning of the Brazilian economy without any reference to the global externalities associated with deforestation and greenhouse gas emissions. However, real opportunity exists for Brazil if an Amazon-wide reduction in deforestation rates were allowed under the international agreement to reduce carbon emissions known as the Kyoto Protocol. A conservation subsidy payment equivalent to a R$1.21 per carbon ton, which is much smaller than the marginal cost of reducing emissions in developed countries, would substantially reduce deforestation rates—by 30 percent—while providing benefits to all regions in Brazil either directly or indirectly. The alternative would be to pursue improvement in technology for perennials, especially labor-intensive technological change, which could reduce deforestation rates considerably.

According to the findings of this report, the Brazilian government, confronted by economic crisis in 1999, has moved in the right direction to contain the damage of the ensuing recession, to attenuate the negative income distribution impact of the crisis, and to limit the adverse environmental effects of the devaluation in terms of deforestation. Nonetheless, the urban sector of the economy suffered considerable income losses as a consequence of the crisis. On the environmental front, logging in the Amazon is likely to increase, while the planned expansion of infrastructure linking the Amazon to the rest of Brazil will increase deforestation for agricultural purposes. However, adoption of a package of provisions combining enforcement of property rights in the Amazon (already under way), technological innovation in perennials within the Amazon and in livestock outside the Amazon (especially in the Northeast), and further provision of subsidies for conservation of forested area in the Amazon would reduce deforestation rates and provide development opportunities. Technological change, while not a policy variable, will affect different producers in different ways and may either increase or reduce deforestation rates. Policymakers should pay attention to what types of technological change are viable and try to facilitate the adoption of those technologies that are compatible with their objectives.

Appendix A

The CGE Model

Table A.1 Definition of parameters and variables in the CGE model

Type	Definition
Sets	
A	Activities
C	Commodities
F	Factors
$FMIG$	Interregionally mobile factors ($FMIG \subset F$)
I	Institutions
HH	Households ($HH \subset I$)
T	Time (years)
Parameters	
α^{TN}	Share of deforestation occurring on tenured land
$\alpha_{f,a}$	CES factor share parameter
α_c^{AC}	Shift parameter for commodity aggregation
α_c^c	Armington function shift parameter
$ashc_{a,c}$	Yield of commod. c per unit of activity a
α_a^d	CES shift parameter
α_c^T	CET function shift parameter
α_c^X	Translog price index constant
$\alpha c_{a,c}^X$	Translog share parameter
$b_{c,a}$	Capital composition matrix
$\beta_{a,c}^X$	Production scale parameter
δ_c	Armington function share parameter
$\delta_{a,c}^{AC}$	Share parameter for commodity aggregation
$dwts_c$	Domestic sales price weights
$econ_c$	Export demand constant
η_c	Export demand price elasticity
γ_c	CET function share parameter
$\gamma_{a,c1,c2}^X$	Transformation parameter between $c1$ and $c2$ for multi-output activity a
$gles_c$	Government consumption shares
$io_{c,a}$	Input-output coefficients
ir	Interest rate

(continued)

Table A.1—Continued

Type	Definition
Parameters *(continued)*	
$itaxa_a$	Rate for indirect tax on activity
$itaxc_c$	Rate for indirect tax on commodity
$kshr_a$	Shares of investment by sector of destination
$make_{a,c}$	Make matrix coefficients
μ_a	Land transformation rate from arable to grassland
μ_g	Land transformation rate from grassland to degraded
ρ_c^C	Armington function exponent
ρ_a^P	CES production function exponent
ρ_c^T	CET function exponent
ρ_c^{AC}	CET function exponent
$sremit_{hh}$	Remittance shares
$strans_{hh}$	Government transfer shares
$syenth_{hh}$	Share of enterprise income to households
$shif_{if}$	Share of factor income to institutions
$sytr_{hh}$	Share of household income transferred to other households
T	Planning horizon for land values
te_c	Tax (+) or subsidy (–) rates on exports
$shtax_{hh}$	Household tax rate
tm_c	Tariff rates on imports
$wfrat_{f_1,f_2}$	Wage ratio: "connected" factor markets
$ymap_{h,hh}$	Household to households map
$zles_c$	Share of investment allocated by commodity
Variables	
$ABSORB$	Total absorption
CD_c	Final demand for private consumption
$DEPREC$	Total depreciation expenditure
DST_C	Inventory investment by sector
DWG_{f_1,f_2}	Wage differential btw f_1 and f_2
EH_{hh}	Household consumption
$ENTSAV$	Enterprise savings
$ENTTAX$	Enterprise tax revenue
ESR	Enterprise savings rate
ETR	Enterprise tax rate
$EXPTAX$	Export subsidy payments
EXR	Exchange rate (R$ per $US)
$FBOR$	Government foreign borrowing
$FDSC_{f,a}$	Factor demand by sector
$FSAV$	Net foreign savings
FS_f	Factor supply
$FXDINV$	Fixed capital investment
$GDPVA$	Value added in market prices
$GDTOT$	Total government consumption
GD_c	Disaggregated government consumption
$GOVGDP$	Government to GDP ratio
$GOVSAV$	Government savings
GR	Government revenue
$HGTR_{hh}$	Government transfers to households
$HREMIT$	Remittances

(continued)

Table A.1—Continued

Type	Definition
Variables *(continued)*	
$HSAVTOT$	Household savings
$HTAX$	Household tax revenue
ID_c	Final investment demand t
$ITAXACT$	Indirect tax revenue from production tax
$ITAXCOM$	Indirect tax revenue from sales taxes
INT_c	Intermediates uses
$INVEST$	Total investment
$INVGDP$	Investment to GDP ratio
MPS_{hh}	Marginal propensity to save by household
PA_a	Domestic activity goods price
PD_c	Domestic commodity goods price
PE_c	Domestic price of exports
$PINDDOM$	Domestic sales price index
PM_c	Domestic price of imports
PQ_c	Price of composite good
PVA_a	Value added price
PWE_c	World price of exports
PWM_c	World price of imports
PX_c	Average output price
$PXAC_{a,c}$	Price of commodity c from activity a
$PXACP_{a,c}$	Pre-tax Price of commodity c from activity a
QA_a	Domestic activity output
QD_c	Domestic Sales
QE_c	Exports
$QFCON_{f_1,f_2}$	Factor conversion from factor f_1 to f_2
$QFMIG_f$	Net migration of factor f
QM_c	Imports
QQ_c	Composite goods supply
QX_c	Domestic commodity output
$QXAC_{a,c}$	Domestic output of commodity c from activity a
$RGDP_c$	Real GDP
$SAC_{a,c}$	Share of commodity c in activity a
$SAVING$	Total savings
$TARIFF$	Tariff revenue
$UESH_f$	Share of factor f going unemployed
WF_f	Average factor price (at base factor demand)
$WFAVG_f$	Average factor price (with current weights)
$WFDIST_{f,a}$	Factor price sectoral proportionality ratios
$YFCTR_f$	Factor income
YH_h	Household income
$YIF_{i,f}$	Share of factor income f to institution i

Note: CES is constant elasticity of substitution. CET is constant elasticity of transformation.

Table A.2 Model equations

	Equation	Description
Price equations		
1.	$PM_c = PWM_c \cdot (1+tm_c) \cdot EXR$	Import price
2.	$PE_c = PWE_c \cdot (1-te_c) \cdot EXR$	Export price
3.	$PQ_c = \frac{(PD_c \cdot QD_c + PM_c \cdot QM_c) \cdot (1+itaxc_c)}{QQ_c}$	Composite commodity prices
4.	$PX_c = \frac{PD_c \cdot QD_c + PE_c \cdot QE_c}{QX_c}$	Producer prices
5.	$PXAC_{a,c} = PXACP_{a,c} \cdot (1+itaxac_{a,c})$	Commodity prices(including indirect taxes)
6.	$\log(PA_a) = a_a^X + \sum_{c \in C} a_{a,c}^X \cdot \log(PXACP_{a,c}) \beta_{a,c_1}^X \cdot \log(QA_a) + 1/2 \cdot \sum_{c_2,c_2 \in C} \gamma_{a,c_1,c_2}^X \cdot \log(PXACP_{a,c1}) \log(PXACP_{a,c2})$	Activity prices(multi-output activities)
7.	$PVA_a = PA_a \cdot (1-itaxa_a) - \sum_{c \in C} PQ_c \cdot io_{c,a}$	Value added prices
8.	$PINDDOM = \sum_c dwts_c \cdot PD_c$	Domestic price index
Quantity equations		
9.	$QA_a = a_a^d \cdot \left[\sum_{f \in F} \alpha_{fa} FDSC_{f,a}^{-\rho_a^P} \right]^{-\frac{1}{\rho_a^P}}$	Activity production (CES)
10.	$FDSC_{f,a} = QA_a \cdot \left[\frac{\alpha_{f,a} \cdot PVA_a}{(a_a^D)^{\rho_a^P} \cdot WF_f \cdot wfdist_{f,a}} \right]^{\sigma_a^P}$	Demand for primary factors
11.	$INT_c = \sum io_{c,a} QA_a$	Intermediate demand
12.	$QXAC_{a,c} = (SAC_{a,c} \cdot PA_a \cdot QA) / PXACP_{a,c}$	Commodity production
13	$SAC_{a,c_1} = ac_{a,c_1}^X + \beta_{a,c_1}^X \cdot \log(QA_a) + \sum_{c_2 \in C} \gamma_{a,c_1,c_2}^X \log(PXACP_{a,c2})$	Value share of commodity c in production by activity a
14.	$QX_c = \alpha_c^{AC} \cdot \left[\sum_{a \in A} \delta_{ac}^{AC} QXAC_{ac}^{-\rho_c^{AC}} \right]^{-\frac{1}{\rho_c^{AC}}}$	Commodity demand(CES aggregation)
15.	$QXAC_{a,c} = QX_c \cdot \left[\frac{\delta_{a,c}^{AC} \cdot PX_c}{(\alpha_a^{AC})^{\rho_a^{AC}} \cdot PXAC_{a,c}} \right]^{\sigma_a^{AC}}$	Disaggregated commodity demand

(continued)

Table A.2—Continued

Equation	Description
Quantity equations (continued)	
16. $QX_c = a_c^T\left[\gamma_c\, QE_c^{\rho_c^T} + (1\ \gamma_c)\, QD_c^{\rho_c^T}\right]^{\frac{1}{\rho_c^T}}$	Output transformation (CET) for exporting sectors
17. $QE_c = QD_c\left[\frac{PE_c\,(1-\gamma_c)}{PD_c\cdot\gamma_c}\right]^{\frac{1}{\rho_c^T-1}}$	Export supply for exports
18. $QQ_c = a_c^C\left[\delta_c\, QM_c^{\rho_c^C} + (1\ \delta_c)\, QD_c^{\rho_c^C}\right]^{\frac{1}{\rho_c^C}}$	Armington assumption:Composite commodity aggregation (CES)
19. $QM_c = QD_c\left[\frac{PD_c\cdot\delta_c}{PM_c(1\ \delta_c)}\right]^{\frac{1}{1+\rho_c^C}}$	Import demand
Income equations	
20. $YFCTR_f = \sum_{a\in A} WF_f\cdot FDSC_{f,a}\cdot WFDIST_{f,a}$	Factor Income
21. $YIF_{i,f} = shif_{i,f}\cdot YFCTR_f$	Income of institution I from factor F
22. $YH_{hh} = \sum_{lb\in LB} YIF_{lb}\cdot shif_{hh,f} + HREMIT_{hh} + \overline{HGTR}_{hh}$	Individual household Income
23. $ITAXCOM = \sum_{c\in C}(PD_c\cdot QD_c + PM_c\cdot QM_c)\cdot itaxc_c$	Commodity taxes
24. $INDTAX = \sum_{a\in A} itaxa_a\cdot PA_a\cdot QA_a$	Activity taxes
25. $TARIFF = \sum_{c\in C} tm_c\cdot PWM_c\cdot QM_c\cdot EXR$	Import tariffs
26. $HTAX = \sum_{hh\in HH} shtax_{hh}\cdot YH_{hh}$	Household taxes
27 $ENTAX = \sum_{ka\in F} sentax\cdot YIF_{"ent",ka}$	Enterprise taxes
28. $GR = HTAX + ENTAX + ITAXCOM + INDTAX + TARIFF$	Government revenue
29. $HSAVTOT = \sum_{hh\in HH} mpssr\cdot MPS_{hh}\cdot YH_{hh}\cdot(1 - shtax_{hh})$	Household savings
30. $ENTSAV = \sum_{ka\in F} esr\cdot YIF_{"ent",ka}\cdot(1 - sentax)$	Enterprise savings
31. $SAVING = HSAVTOT + ENTSAV + GOVSAV + FSAV\cdot EXR$	Total savings

(continued)

Table A.2—Continued

	Equation	Description
Expenditure equations		
32.	$GR = \sum_{c \in C} PQ_c \cdot CD_c + GOVSAV + \sum_{hh \in HH} \overline{HGTR}_{hh}$	Government expenditure
33.	$EH_{hh} = (1 - shtax_{hh}) \cdot (1 - MPS_{hh} \cdot mpssr) \cdot YH_{hh}$	Household consumption expenditures
34.	$PQ_c \cdot CD_c = \sum_{hh \in HH} cles_{c,hh} \cdot EH_{hh}$	Household consumption demand
35.	$GD_c = gles_c \cdot GDTOT$	Government consumption demand
36.	$ID_c = zles_c \cdot FXDINV$	Fixed investment demand
37.	$INVEST = \sum_{c \in C} PQ_c \cdot (ID_c + DST_c)$	Investment value
Factor supply and demand, and migration relationships		
38a.	$FS_{f1} = \overline{FS}_{f1} + \sum_{a \in A} \mu_{a,f1,f2} \cdot FDSC_{a,f2} + \sum_{f2 \in f} QFCON_{f1,f2}$	Factor supply (no migration)Includes factor transformation for physical causes and factor conversion (such as deforestation)
38b.	$FS_f = \overline{FS}_f + QFMIG_f$	Factor supply (with migration)
39.	$QFMIG_{f_1} = \sum_{f_2} (OUTMIG_{f_2,f_1} - OUTMIG_{f_1,f_2})$	Net migration arriving into f1
40.	$WFAVG_f = \sum_{a \in A} WF_f \cdot FDSC_{a,f} \cdot WFDIST_{a,f} \Big/ \sum_{a \in A} FDSC_{a,f}$	Average factor wage
41.	$WFAVG_{f1} = wfrat_{f1,f2} \cdot (1 + DWG_{f_1,f_2}) \cdot WFAVG_{f2}$	For "connected" factor markets the wage ratio is constrained
42.	$DWG_{f1,f2} > dwt_{f1,f2} \quad \left[OUTMIG_{f1,f2} > 0 \right]$	Migration occurs when wage differential exceeds threshold
43a.	$\sum_{f \in FMIG} QFMIG_f = 0$	Conservation of total factor supply (for factors that are "connected" through migration or conversion)
43b.	$\sum_{f_1,f_2 \in FCON} QFCON_{f_1,f_2} = 0$	Conservation of total factor supply (for factors that are "connected" through migration or conversion)
44a.	$FS_f = \sum_{a \in A} FDSC_{a,f}$	Factor market equilibrium (fully employed factors)

(continued)

Table A.2—Continued

Equation		Description
Factor supply and demand, and migration relationships (continued)		
44b.	$FS_f > \sum_{a \in A} FDSC_{a,f} \left[WF_f > wf_f^{\min} \right]$	Factor market equilibrium (potentially unemployed factors)
45.	$UESH_f = \left[FS_f - \sum_{a \in A} FDSC_{a,f} \right] \Big/ FS_f$	Share of factor going unemployed (potentially unemployed factors)
46.	$PX_{"def"} = \frac{WFAVG_{"ar"}}{i+\mu_a}[1 - e^{-(i+\mu_a)T}] + \frac{WFAVG_{"gr"}}{i+\mu_g}[1 - e^{-(i+\mu_g)T}] - \frac{WFAVG_{"gr"}}{i+\mu_a+\mu_g}[1 - e^{-(i+\mu_a+\mu_g)T}] - \alpha \frac{WFAVG_{"for"}}{i}$	Deforestation demand: price is the expected NPV of returns to land
Macroeconomic closures		
	$QQ_c = INT_c + CD_c + GD_c + ID_c + DST_c$	Commodity market equilibrium
47.	$\sum_{c \in CM} PM_c \cdot QM_c = \sum_{c \in CE} PE_c \cdot QE_c + FSAV + \sum_{hh \in HH} HREMIT_{hh}$	External account balance
48.	$ABSORB = \sum_{c \in C} PQ_c (CD_c + ID_c + GD_c + DST_c)$	Total absorption
49a.	$\overline{GOVABS} = \frac{\sum_{c \in C} PQ_c \cdot GD_c}{ABSORB}$	Government consumption and investment demand (fixed share of absorption
49b.	$\overline{INVABS} = \frac{\sum_{c \in C} PQ_c \cdot ID_c}{ABSORB}$	Government consumption and investment demand (fixed share of absorption)
50.	$SAVING = INVEST$	Saving-investment balance

Notes: The mathematical model statement is divided into the following sections: price equations, quantity equations, income equations, expenditure equations, factor market relationships, and macroeconomics closure.

APPENDIX B

Structure of the Economy

Table B.1 Regional production by commodity, 1995 (R$ billion)

	Small Farms				Large Farms			
Product	Amazon	Northeast	Center-West	South/Southeast	Amazon	Northeast	Center-West	South/Souhest
Annuals								
Maize	0.090	0.337	0.078	2.407	0.036	0.210	0.928	2.150
Rice	0.145	0.262	0.031	0.500	0.115	0.099	0.203	1.751
Beans	0.097	0.609	0.018	0.735	0.010	0.146	0.080	0.395
Manioc	0.917	0.466	0.038	0.857	0.034	0.059	0.035	0.122
Other annuals	0.094	0.523	0.091	3.995	0.030	0.412	0.421	2.080
Sugar	0.014	0.166	0.015	0.901	0.012	1.378	0.675	5.326
Soy	0.000	0.000	0.020	0.894	0.001	0.158	1.322	1.438
Horticultural products	0.090	0.227	0.067	1.392	0.005	0.040	0.013	0.146
Perennials								
Coffee	0.140	0.111	0.015	2.421	0.011	0.097	0.010	2.221
Cocoa	0.032	0.271	0.001	0.005	0.010	0.196	0.001	0.01
Other perennial	0.444	1.037	0.054	2.443	0.106	0.579	0.076	1.829
Animal Products								
Milk	0.466	0.738	0.403	4.115	0.218	0.634	1.064	2.831
Cattle and swine	1.027	1.641	0.405	3.477	1.646	1.412	3.642	4.332
Poultry	0.29	0.783	0.181	4.264	0.069	0.241	0.136	1.067
Other agriculture	0.488	1.329	0.378	0.523	0.126	0.502	0.962	0.16
Forest products[a]								
Forest extraction	0.194	0.135	0.003	0.082				
Logging	0.563	0.385	0.175	3.088				
Deforestation	0.603	...	...	...				

Source: IBGE 1998a.

Notes: [a]For forest products, figures for small and large farms are combined.

Table B.2 Factor Intensities for Amazon agriculture in the base (in terms of output)

	Small farms				Large farms				
Factor	Annual	Perennial	Animal	Other agriculture	Annual	Perennial	Animal	Other agriculture	All forest products
Annuals									
Labor	546	354	277	226	336	483	25	291	232
Capital	209	431	459	325	159	421	638	355	205
Arable land	717	456		179	1,168	469		139	...
Grassland	...	...	3,268	...	...	...	9,182	...	...
Forest land	...	...	...	...	...	...	...	...	21,634

Notes: The units express: for labor, the number of workers involved in producing R$ 1 million in the activity; for capital, the monetary equivalent in thousands of R$ of physical capital involved in producing for R$ 1 million; for land, the number of hectares required to produce R$ 1 million worth of output. Only cells in the same row can be compared because the units of measure are different.

APPENDIX C

Elasticities and Sensitivity Analysis

The The elasticities used in the model are presented in Table A5. These include trade substitution elasticities which describe, on the demand side, the degree to which imported and domestically produced goods are substitutes in consumption, and, on the production side, the extent to which goods produced for export or for the domestic market are interchangeable in the production process. Trade substitution elasticities are critical when the simulations to be performed affect the price of traded goods relative to non-traded goods (for example, a devaluation). For the technological innovation scenarios, changing these elasticities does not change the results.

Factor substitution elasticities were obtained from partial equilibrium estimates and adapted for agriculture to the different regions in Brazil and to the different types of producers. The underlying rationale is that large farms can substitute more easily between factors, and that Amazon production has a more restricted set of substitution possibilities relative to the other regions. Here too, the elasticity values do not affect our deforestation estimates in a significant way (a ± 30% change in the elasticities leads to a ± 4% change in the deforestation results, and does not affect the policy implications of the results described in the paper).

The elasticities that really make a difference to our analysis are the ones relating to the access of regional products into the national commodity market. These were assumed to be high for all agricultural products, implying that Amazon products could easily be absorbed by the Brazilian domestic and export markets. When the simulations were performed with lower elasticities (elas.= 2 for all agricultural products) the deforestation results changed considerably.

With lower absorption by the national commodity market, the highest increase in deforestation rates was a 15% increase when land-saving, sustainability-improving, technological innovation in annuals is adopted. The lower elasticity dampens the results presented in the body of the report because the terms of trade are more sensitive to increased production associated with technological improvement in the Amazon.

Table C.1 Model elasticities

Source	Elasticity
	Import substitution elasticities
Agricultural goods	2.0 to 4.0
Processed foods	1.25
Mining and oil	1.15
Industry	1.50
Construction	0.50
Trade and transportation	0.75
Services	0.65
	Export transformation elasticities
Agricultural goods	1.1 to1.5
Processed foods	3.00
Mining and oil	1.75
Industry	2.25
Construction	0.50
Trade and transportation	0.75
Services	0.65
	Factor substitution elasticities
Amazon large-farm activities	0.80
Amazon small-farmsactivities	0.40
Large-farm activities (other goods)	1.50
Small-farm activities (other goods)	0.90
Forest product activities	0.50
Mining and oil	0.50
Industry	0.50
Construction	1.50
Trade and transportation	0.90
Services	1.50
	Substitution elasticity of regional goods in the national commodity market
Agricultural goods	12.00
Nonagricultural goods	—no regionality—

APPENDIX D

Results for Devaluation Scenarios

Table D.1 Short-run changes in production with currency devaluation, balanced adjustment scenario (%)

Production Activity	Base value (R$ billion)	Real devaluation of Brazilian currency			
		10%	20%	30%	40%
Agriculture					
Coffee	5.03	0.23	0.72	1.46	2.38
Cocoa	0.52	−1.04	−2.66	−4.75	−7.31
Maize	6.24	0.09	0.21	0.43	0.71
Rice	3.11	0.69	1.58	2.68	3.92
Beans	2.09	−1.04	−2.40	−3.96	−5.68
Manioc	2.53	−2.07	−4.25	−6.5	−8.79
Other perennials	6.57	−0.68	−1.82	−3.23	−4.90
Other annuals	7.65	3.01	5.34	7.24	8.76
Sugar	8.49	0.90	2.04	3.43	5.06
Soy	3.83	1.20	2.64	4.29	6.08
Horticulture	1.98	1.58	3.27	5.13	7.16
Milk	10.47	−1.30	−2.57	−3.74	−4.84
Cattle and swine	17.58	0.50	1.19	2.08	3.12
Poultry	7.03	−1.50	−3.04	−4.54	−6.01
Forest extraction	0.41	0.95	−0.35	−3.56	−8.50
Logging	4.21	1.51	3.06	4.68	6.40
Deforestation	0.60	−2.13	−4.24	−6.51	−9.16
Other agriculture	4.47	−0.34	−1.85	−3.94	−6.48
Nonagricultural sectors					
Processed. food	147.49	0.27	0.83	1.66	2.68
Oil and mining	35.75	2.49	5.30	8.48	12.06
Manufacturing	373.13	1.40	3.01	4.86	6.99
Construction	102.8	−3.25	−6.71	−10.43	−14.51
Trade and transportation	118.69	1.35	2.89	4.65	6.65
Services	397.89	−2.38	−5.01	−7.95	−11.24

Table D.2 Short-run effects of currency devaluation on per capita income, balanced-adjustment scenario

Household type	Base value (R$/year)	Real devaluation of Brazilian currency 10%	20%	30%	40%
Urban low	1,349	–1.60	–3.1	–4.49	–5.78
Urban medium	2,548	–1.25	–2.37	–3.35	–4.2
Rural low	295	2.43	5.71	9.75	14.59
Rural medium	700	1.85	4.42	7.64	11.52
High	5,511	–0.67	–1.12	–1.32	–1.30

Table D.3 Changes in macroeconomic aggregates with currency devaluation balanced-adjustment scenario (%)

Balanced-adjustment scenario	Base value (R$ billion)	Real devaluation of Brazilian currency 10%	20%	30%	40%
GDP	658.14	–1.03	–2.34	–3.93	–5.85
Consumption	429.75	–3.77	–7.78	–12.1	–16.77
Investment	126.64	–3.45	–7.1	–11.04	–15.35
Government consumption	110.49	–3.17	–6.68	–10.58	–14.92
Exports	46.31	22.12	46.77	73.91	103.44
Imports	–55.04	–12.84	–23.29	–31.99	–39.36

Table D.4 Long-run changes in production with currency devaluation, balanced-adjustment scenario (%)

Production activity	Base value (R$ billion)	Real devaluation of Brazilian currency 10%	20%	30%	40%
Agriculture					
Coffee	5.03	1.07	2.67	4.84	7.56
Cocoa	0.52	0.71	1.10	1.18	0.85
Maize	6.24	0.94	2.14	3.66	5.52
Rice	3.11	1.53	3.51	6.03	9.06
Beans	2.09	0.10	–0.04	–0.33	–0.75
Manioc	2.53	–0.72	–1.59	–2.49	–3.46
Other perennials	6.57	0.68	1.07	1.28	1.27
Other annuals	7.65	4.29	8.24	12.06	15.76
Sugar	8.49	1.30	2.92	4.89	7.22
Soy	3.83	2.08	4.73	8.01	11.88
Horticulture	1.98	1.97	4.12	6.05	9.11
Milk	10.47	–0.30	–0.40	–0.26	0.12
Cattle and swine	17.58	1.32	3.09	5.35	8.11
Poultry	7.03	–0.38	–0.63	–0.7	–0.6
Forest extraction	0.41	2.45	2.89	1.47	–1.77
Logging	4.21	1.76	3.57	5.44	7.40
Deforestation	0.60	0.81	0.77	1.09	1.77
Other agriculture	4.47	1.39	1.78	1.60	0.95

(continued)

Table D.4—Continued

Production activity	Base value (R$ billion)	Real devaluation of Brazilian currency 10%	20%	30%	40%
Nonagriculture sectors					
Processed food	147.49	1.08	2.71	4.92	7.68
Oil and mining	35.75	2.50	5.23	8.20	11.45
Manufacturing	373.13	1.48	3.11	4.90	6.90
Construction	102.80	–3.13	–6.50	–10.23	–14.37
Trade and transportation	118.69	1.73	3.70	5.93	8.44
Services	397.89	–2.21	–4.77	–7.76	–11.22

Table D.5 Long-run effects of currency devaluation on per capita income, balanced-adjustment scenario (%)

Household type	Base value (R$/year)	Real devaluation of Brazilian currency 10%	20%	30%	40%
Urban low	1,349	1.01	2.41	4.33	6.85
Urban medium	2,548	0.51	1.31	2.45	3.97
Rural low	295	–1.32	–2.2	–2.74	–2.9
Rural medium	700	–1.75	–3.12	–4.23	–5.04
High	5,511	–0.08	0.1	0.55	1.27

Table D.6 Long-run changes in macroeconomic aggregates with currency devaluation, balanced-adjustment scenario (%)

Balanced-adjustment scenario	Base value (R$ billion)	Real devaluation of Brazilian currency 10%	20%	30%	40%
GDP	658.14	–0.77	–1.85	–3.28	–5.10
Consumption	429.75	–3.47	–7.23	–11.39	–16.02
Investment	126.64	–3.33	–6.91	–10.85	–15.22
Government consumption	110.49	–3.02	–6.47	–10.46	–15.04
Exports	46.31	22.64	48.18	76.71	108.24
Imports	–55.04	–12.58	–22.85	–31.45	–38.79

Table D.7 Short-run changes in production with currency devaluation, capital-flight scenario (%)

Production activity	Base value (R$ billion)	Real devaluation of Brazilian currency 10%	20%	30%	40%
Agriculture					
Coffee	5.03	0.84	1.76	2.71	3.6
Cocoa	0.52	−1.82	−4.14	−6.75	−9.54
Maize	6.24	0.62	1.2	1.79	2.32
Rice	3.11	1.29	2.6	3.9	5.1
Beans	2.09	−0.87	−1.9	−2.96	−4
Manioc	2.53	−1.72	−3.4	−4.98	−6.42
Other perennials	6.57	−1.01	−2.32	−3.72	−5.13
Other annuals	7.65	2.12	3.52	4.43	4.95
Sugar	8.49	1.42	3.07	4.99	7.18
Soy	3.83	1.6	3.18	4.69	6.03
Horticulture	1.98	1.79	3.71	5.84	8.24
Milk	10.47	−1	−1.93	−2.75	−3.48
Cattle and swine	17.58	0.97	1.96	2.97	3.91
Poultry	7.03	−1.37	−2.72	−3.98	−5.17
Forest extraction	0.41	−2.51	−7.46	−14.35	−22.83
Logging	4.21	1.65	3.36	5.22	7.26
Deforestation	0.60	2.11	4.15	5.59	5.79
Other agriculture	4.47	−1.63	−4.11	−6.87	−9.78
Nonagriculture sectors					
Processed food	147.49	0.93	1.94	3	3.99
Oil and mining	35.75	−0.65	−1.56	−2.73	−4.14
Manufacturing	373.13	1.83	4	6.6	9.75
Construction	102.8	−12.22	−26.29	−42.63	−61.73
Trade and transportation	118.69	1.65	3.5	5.62	8.06
Services	397.89	−0.44	−0.78	−1.02	−1.15

Table D.8 Short-run effects of currency devaluation on per capita income, capital-flight scenario (%)

Household type	Base value (R$/year)	Real devaluation of Brazilian currency 10%	20%	30%	40%
Urban low	1,349	−1.68	−3.24	−4.67	−5.94
Urban medium	2,548	−1.37	−2.62	−3.71	−4.62
Rural low	295	5.77	12.82	21.12	30.67
Rural medium	700	4.45	9.95	16.48	24.03
High	5,511	−0.69	−1.14	−1.32	−1.22

Table D.9 Short-run changes in macroeconomic aggregates with currency devaluation, capital-flight scenario (%)

Capital-flight scenario	Base value (R$ billion)	Real devaluation of Brazilian currency 10%	20%	30%	40%
GDP	658.14	–0.88	–2.02	–3.44	–5.19
Consumption	429.75	–0.64	–0.96	–0.94	–0.52
Investment	126.64	–13.51	–29.07	–47.15	–68.31
Government consumption	110.49	–2.73	–5.71	–8.99	–12.61
Exports	46.31	21.78	45.92	72.53	101.73
Imports	–55.04	–12.72	–23.1	–31.71	–38.96

Table D.10 Long-run changes in production with currency devaluation, capital-flight scenario (%)

Production Activity	Base value (R$ billion)	Real devaluation of Brazilian currency 10%	20%	30%	40%
Agriculture					
Coffee	5.03	2.25	5.00	8.20	11.68
Cocoa	0.52	1.31	2.33	3.08	3.50
Maize	6.24	2.07	4.45	7.12	9.98
Rice	3.11	2.69	5.82	9.36	13.17
Beans	2.09	1.10	2.10	3.10	4.08
Manioc	2.53	0.57	1.13	1.78	2.47
Other perennials	6.57	1.46	2.73	3.89	4.92
Other annuals	7.65	4.53	8.68	12.59	16.19
Sugar	8.49	1.98	4.32	7.02	10.03
Soy	3.83	3.11	6.70	10.73	14.99
Horticulture	1.98	2.36	4.92	7.70	10.70
Milk	10.47	0.74	1.75	3.01	4.44
Cattle and swine	17.58	2.37	5.17	8.35	11.78
Poultry	7.03	0.61	1.39	2.36	3.44
Forest extraction	0.41	0.82	–0.83	–4.76	–10.72
Logging	4.21	1.99	4.03	6.17	8.40
Deforestation	0.60	5.37	10.25	15.37	20.24
Other agriculture	4.47	1.70	2.59	3.06	3.20
Nonagriculture sectors					
Processed food	147.49	2.26	5.03	8.25	11.75
Oil and mining	35.75	–0.54	–1.44	–2.72	–4.36
Manufacturing	373.13	1.78	3.79	6.09	8.73
Construction	102.8	–12.12	–26.33	–43.13	–62.93
Trade and transportation	118.69	2.15	4.55	7.24	10.19
Services	397.89	–0.88	–1.85	–2.94	–4.11

Table D.11 Long-run effects of currency devaluation on per capital income, capital-flight scenario (%)

Household type	Base value (R$/year)	Real devaluation of Brazilian currency 10%	20%	30%	40%
Urban low	1,349	2.48	5.63	9.57	14.31
Urban medium	2,548	1.29	2.96	5.06	7.52
Rural low	295	–1.35	–1.96	–1.84	–0.92
Rural medium	700	–2.26	–3.87	–4.86	–5.12
High	5,511	–0.03	0.2	0.69	1.45

Table D.12 Long-run changes in macroeconomic aggregates with currency devaluation, capital-flight scenario (%)

Capital-flight scenario	Base value (R$ billion)	Real devaluation of Brazilian currency 10%	20%	30%	40%
GDP	658.14	–0.82	–2.01	–3.60	–5.67
Consumption	429.75	–0.57	–0.90	–1.01	–0.86
Investment	126.64	–13.37	–29.05	–47.58	–69.43
Government consumption	110.49	–3.27	–6.99	–11.23	–15.97
Exports	46.31	22.96	48.87	77.78	109.48
Imports	–55.04	–12.70	–23.13	–31.88	–39.32

APPENDIX E

Results for Transportation Cost Reduction Scenarios

Table E.1 Short-run changes in production with reduction in transportation costs

Production Activity	Base value (R$ billion)	Reduction in transportation costs for Amazon products 5%	10%	15%	20%
Agriculture					
Coffee	5.03	–0.02	–0.02	–0.03	–0.04
Cocoa	0.52	0.02	0.03	0.04	0.05
Maize	6.24	0.00	0.00	–0.01	–0.01
Rice	3.11	0.00	0.00	0.00	–0.01
Beans	2.09	–0.03	–0.03	–0.03	–0.03
Manioc	2.53	0.02	0.08	0.14	0.17
Other perennials	6.57	0.00	0.00	0.00	0.00
Other annuals	7.65	0.06	0.07	0.08	0.09
Sugar	8.49	0.00	0.00	0.00	0.00
Soy	3.83	0.01	0.01	0.01	0.01
Horticulture	1.98	0.00	0.00	–0.01	–0.01
Milk	10.47	–0.01	–0.02	–0.02	–0.02
Cattle and swine	17.58	–0.01	–0.02	–0.03	–0.04
Poultry	7.03	–0.02	–0.02	–0.03	–0.03
Forest extraction	0.41	0.36	0.66	0.94	1.19
Logging	4.21	0.01	0.02	0.03	0.04
Deforestation	0.60	4.25	8.04	11.53	14.71
Other agriculture	4.47	0.21	0.23	0.25	0.27
Nonagriculture sectors					
Processed food	147.49	0.00	0.00	0.00	0.00
Oil and mining	35.75	0.00	0.00	0.00	0.00
Manufacturing	373.13	0.00	0.00	0.00	0.00
Construction	102.80	0.00	0.00	0.00	0.00
Trade and transportation	118.69	–0.09	–0.18	–0.26	–0.35
Services	397.89	0.01	0.02	0.03	0.05

Table E.2 Short-run effects of reduction in transportation costs on per capita income (%)

Household type	Base value (R$/year)	Reduction in transportation costs for Amazon products 5%	10%	15%	20%
Urban low	1,349	0.00	0.01	0.01	0.01
Urban medium	2,548	0.00	0.00	0.00	0.00
Rural low	295	0.28	0.55	0.83	1.11
Rural medium	700	0.21	0.41	0.62	0.83
High	5,511	0.01	0.02	0.04	0.05

Table E.3 Short-run changes in macroeconomic aggregates with reduction in transportation cost (%)

Activity	Base value (R$ billion)	Reduction in transportation costs for Amazon products 5%	10%	15%	20%
GDP	658.14	0.02	0.03	0.04	0.06
Consumption	429.75	0.02	0.04	0.05	0.07
Investment	126.64	0.00	0.00	0.00	0.00
Government consumption	110.49	0.02	0.04	0.06	0.08
Exports	46.31	–0.01	–0.02	–0.03	–0.05
Imports	–55.04	–0.01	–0.02	–0.03	–0.04

Table E.4 Long-run changes in production with reduction in transportation costs (%)

Production Activity	Base value (R$ billion)	Reduction in transportation costs for Amazon products 5%	10%	15%	20%
Agriculture					
Coffee	5.03	0.07	0.11	0.14	0.17
Cocoa	0.52	0.40	0.69	0.99	1.28
Maize	6.24	0.08	0.11	0.14	0.16
Rice	3.11	0.12	0.20	0.27	0.34
Beans	2.09	0.06	0.06	0.06	0.06
Manioc	2.53	0.92	1.73	2.57	3.42
Other perennials	6.57	0.30	0.50	0.69	0.87
Other annuals	7.65	0.15	0.18	0.19	0.20
Sugar	8.49	0.03	0.05	0.07	0.08
Soy	3.83	0.10	0.15	0.19	0.23
Horticulture	1.98	0.02	0.02	0.02	0.01
Milk	10.47	0.11	0.16	0.21	0.25
Cattle and swine	17.58	0.08	0.12	0.15	0.18
Poultry	7.03	0.13	0.18	0.22	0.26
Forest extraction	0.41	1.00	1.79	2.55	3.30
Logging	4.21	0.04	0.06	0.09	0.11
Deforestation	0.60	12.64	22.82	31.43	38.42
Other Agriculture	4.47	0.50	0.69	0.85	0.98

(continued)

Table E.4—Continued

Production Activity	Base value (R$ billion)	Reduction in transportation costs for Amazon products 5%	10%	15%	20%
Nonagriculture sectors					
Processed food	147.49	0.10	0.16	0.22	0.27
Oil and mining	35.75	–0.03	–0.03	–0.04	–0.05
Manufacturing	373.13	–0.02	–0.03	–0.04	–0.05
Construction	102.80	–0.02	–0.02	–0.02	–0.02
Trade and transportation	118.69	–0.04	–0.10	–0.18	–0.26
Services	397.89	–0.04	–0.05	–0.07	–0.09

Table E.5 Long-run effects of reduction in transportation costs on per capita income (%)

Household type	Base value (R$/year)	Reduction in transportation costs for Amazon products 5%	10%	15%	20%
Urban low	1,349	0.13	0.21	0.31	0.42
Urban medium	2,548	0.09	0.15	0.22	0.29
Rural low	295	–0.01	0.09	0.19	0.29
Rural medium	700	–0.04	0.02	0.07	0.12
High	5,511	0.03	0.06	0.09	0.11

Table E.6 Long-run changes in macroeconomic aggregates with reduction in transportation costs (%)

Activity	Base value (R$ billion)	Reduction in transportation costs for Amazon products 5%	10%	15%	20%
GDP	658.14	0.02	0.03	0.05	0.06
Consumption	429.75	0.03	0.05	0.07	0.09
Investment	126.64	–0.01	–0.01	–0.01	–0.01
Government consumption	110.49	–0.04	–0.05	–0.07	–0.09
Exports	46.31	0.07	0.11	0.14	0.17
Imports	–55.04	–0.03	–0.05	–0.07	–0.09

Appendix F

Results for Technological Change Scenarios

Table F.1 Wage impact of technological change in the Amazon (% change)

		TFPI		TFP2		TFP3		TFP5	TFP7
		SR	LR	SR	LR	SR	LR	LR	LR
TFP Annuals production	Unskilled lab	6.2	0.1	18.0	0.2	29.8	–0.9	–6.7	–13.8
	Smallhold. K	–5.2	–0.7	–15.0	–2.3	–23.8	–4.7	–11.3	–17.6
	Large Farm K	–1.1	–0.7	–2.6	–2.3	–3.4	–4.7	–11.3	–17.6
	Arable land	18.6	21.9	56.7	63.5	101.1	118.8	184.2	140.1
	Pasture	–1.0	–0.3	–3.5	–1.2	–7.1	–2.4	–6.3	–10.6
TFP Perennials production	Unskilled lab	1.2	0.2	4.3	1.1	8.4	1.7	–0.3	–6.3
	Smallhold. K	3.2	–0.1	7.2	–0.4	11.2	–1.2	–4.6	–9.9
	Large Farm K	0.4	–0.1	1.3	–0.4	3.3	–1.2	–4.6	–9.9
	Arable land	3.2	6.4	9.3	16.8	16.7	32.1	68.9	96.6
	Pasture	–2.8	–1.5	–8.3	–4.4	–15.0	–7.9	–14.9	–19.8
TFP Animal products	Unskilled lab	...	–1.9	...	–4.7	...	–8.8	–19.9	–31.2
	Smallhold. K	9.8	–0.9	29.6	–2.7	54.1	–5.8	–15.3	–27.0
	Large Farm K	7.5	–0.9	22.1	–2.7	38.5	–5.8	–15.3	–27.0
	Arable land	...	1.4	2.7	...	6.3	17.3	23.4	
	Pasture	15.0	14.8	45.0	42.9	78.7	79.9	145.3	143.2
Labor intensitve Annuals production	Unskilled lab	13.1	0.1	43.2	0.4	84.1	0.3	2.7	8.1
	Smallhold. K	–11.3	–1.2	–32.7	–3.6	–53.0	–6.5	–11.0	–13.1
	Large Farm K	–2.3	–1.2	–6.2	–3.6	–9.1	–6.5	–11.0	–13.1
	Arable land	7.5	21.4	11.2	57.0	7.1	89.1	122.1	138.7
	Pasture	–1.8	–0.5	–5.8	–1.7	–10.8	–2.8	–3.9	–3.5
Labor intensive Perennials production	Unskilled lab	4.1	0.6	16.3	2.7	36.4	5.7	13.9	24.5
	Smallhold. K	0.6	–0.2	–5.8	–0.6	–19.2	–0.8	–0.6	0.6
	Large Farm K	–1.0	–0.2	–3.7	–0.6	–6.4	–0.8	–0.6	0.6
	Arable land	...	5.4	...	11.2	...	15.1	15.1	9.2
	Pasture	–4.0	–2.3	–10.9	–6.7	–17.9	–11.4	–19.1	–24.6
Capital intensive Annuals production	Unskilled lab	2.2	0.2	2.1	1.5		2.5	4.3	6.1
	Smallhold. K	5.6	–0.2	23.6	0.4	53.8	1.7	5.6	11.2
	Large Farm k	4.0	–0.2	16.6	0.4	39.6	1.7	5.6	11.2
	Arable land	20.3	20.2	46.2	46.3	69.6	68.6	103.0	131.8
	Pasture	–3.2	–0.3	–10.1	–0.3	–19.9	–0.1	0.5	1.4

(continued)

Table F.1—Continued

		TFPI		TFP2		TFP3		TFP5	TFP7
		SR	LR	SR	LR	SR	LR	LR	LR
Capital intensive Perennials production	Unskilled lab	...	0.2	...	1.3	...	2.7	5.3	8.1
	Smallhold. K	9.6	0.1	30.4	0.7	60.2	2.0	6.0	12.4
	Large Farm k	3.4	0.1	15.1	0.7	37.3	2.0	6.0	12.4
	Arable land	4.3	7.5	7.8	16.7	7.3	24.4	34.6	42.0
	Pasture	-4.5	-1.7	-12.9	-4.1	-23.3	-5.8	-7.8	-8.8
Capital intensive Animal products	Unskilled lab	...	-2.2		-4.5	...	-6.5	-7.5	-5.0
	Smallhold. K	38.9	0.6	153.8	4.1	353.0	11.5	39.3	86.9
	Large Farm k	38.8	0.6	142.8	4.1	309.1	11.5	39.3	86.9
	Large Farm k	38.8	0.6	142.8	4.1	309.1	11.5	39.3	86.9
	Arable land	...	3.5	...	11.9	...	26.7	68.6	121.8
	Pasture	1.3	15.6	0.9	41.2	-1.1	67.0	113.6	157.3
Capital and labor intensive Annuals production	Unskilled lab	10.8	0.0	31.0	-0.8	47.3	-3.8	-11.7	-15.4
	Smallhold. K	-8.2	-1.2	-22.0	-4.0	-31.9	-8.1	-16.5	-21.0
	Large farm K	-1.3	-1.2	-2.3	-4.0	-2.6	-8.1	-16.5	-21.0
	Arable land	10.2	21.5	14.6	52.9	1.3	62.1	8.0	
	Pasture	-2.2	-0.6	-7.2	-2.2	-12.6	-4.5	-9.9	-13.7
Capital and labor intensive Perennials production	Unskilled lab	2.3	0.4	8.1	1.4	13.8	1.0	-4.2	-9.4
	Smallhold. K	4.6	-0.2	10.7	-0.9	17.0	-2.6	-7.9	-12.1
	Large farm K	0.7	-0.2	3.3	-0.9	6.5	-2.6	-7.9	-12.1
	Arable land	0.5	6.7	...	16.2	...	24.3	30.1	28.6
	Pasture	-4.6	-2.3	-13.9	-6.9	-22.9	-12.1	-20.0	-24.3
Capital and labor intensive Animal products	Unskilled lab	...	-3.3	...	-7.2	9.4	-10.0	-13.0	-13.8
	Smallhold. K	29.7	-0.5	118.8	0.0	271.6	3.8	22.3	54.4
	Large farm K	28.2	-0.5	98.5	0.0	208.5	3.8	22.3	54.4
	Arable land	...	5.6	...	15.3	...	30.6	84.3	170.4
	Pasture	5.2	16.1	7.8	43.6	5.8	74.6	134.3	183.6
Capital and labor intensive + sustain Annuals production	Unskilled lab	10.9	-0.1	31.2	-1.0	47.3	-4.0	-12.0	-15.4
	Smallhold. K	-8.2	-1.2	-22.0	-4.0	-31.9	-8.2	-16.6	-21.0
	Large farm K	-1.2	-1.2	-2.3	-4.0	-2.5	-8.2	-16.6	-21.0
	Arable land	6.8	18.2	7.2	43.8	...	46.5	...	...
	Pasture	-2.2	-0.6	-7.2	-2.1	-12.7	-4.5	-10.0	-13.7
Capital and labor intensive + sustain. Animal Products	Unskilled lab	...	-3.4	...	-7.4	9.7	-10.1	-12.1	-10.6
	Smallhold. K	30.3	-0.6	121.3	-0.1	276.9	3.7	22.6	55.5
	Large farm K	28.8	-0.6	100.5	-0.1	212.5	3.7	22.6	55.5
	Arable land	...	5.9	...	15.8	...	30.9	81.0	154.8
	Pasture	4.2	15.5	5.4	41.8	1.8	70.7	124.7	166.4

Notes: TFP is total factor productivity, SR is short run, and LR is long run. Leaders (...) indicate a nil or negligible amount.

Table F.2 Short-run impact of technological change on different producer types (%)

	Smallholders						Farm enterprises					
	Production			Terms of trade			Production			Terms of trade		
Simulations	TFP1	TFP2	TFP3	TFP1	TFP2	TFP3	TFP1	TFP2	TFP3	TFP1	TFP2	TFP3
TFP_AN	4.4	16.7	39.4	–0.6	–4.0	–10.5	1.7	10.2	30.9	–1.9	–5.3	–9.3
TFP_PE	1.9	8.6	24.8	0.6	0.8	–0.8	0.5	4.0	14.0	–0.6	–2.7	–7.2
TFP_LV	6.4	25.8	63.4	–2.2	–7.9	–16.8	10.3	36.1	79.4	0.0	–3.4	–12.3
LABIN_AN	7.0	24.0	45.7	–1.0	–6.0	–13.0	2.9	14.1	30.7	–2.9	–7.5	–12.0
LABIN_PE	2.5	6.2	9.1	1.0	1.1	–0.8	0.0	1.3	3.9	–1.2	–2.7	–3.3
CAPIN_AN	–0.2	–5.6	–11.6	–2.4	–6.5	–10.2	2.9	5.3	5.0	1.7	6.4	10.2
CAPIN_PE	0.1	–5.4	–13.7	–0.5	–3.9	–8.4	2.1	7.8	13.2	1.5	8.2	16.2
CAPIN_LV	–0.5	–6.8	–12.6	–3.6	–13.0	–22.1	8.4	26.9	47.5	2.7	13.6	25.2
LBCAP_AN	7.6	31.0	67.7	–1.5	–8.7	–19.5	4.2	24.9	68.3	–2.5	–7.0	–11.4
LBCAP_PE	3.5	18.6	52.6	0.7	–0.4	–4.9	1.3	11.0	32.6	–1.0	–5.1	–13.7
LBCAP_LV	9.2	26.4	36.5	–3.6	–13.8	–25.1	14.3	47.0	86.5	–1.1	–2.8	–0.8
DGLBK_AN	7.6	31.2	67.8	–1.5	–8.8	–19.5	4.3	25.3	68.4	–2.5	–6.9	–11.3
DGLBK_LV	9.3	27.4	38.8	–3.8	–14.3	–25.9	14.9	49.0	91.0	–1.0	–2.8	–1.3

APPENDIX G

Results for Non-Amazon Technological Change Scenarios

Table G.1 Change in per capita regional agricultural income for non-Amazon technological change: Decomposing the impact of innovation on producers by type of activity

		Improvement in annuals				Improvement in Perennials			
Region	Producer	All except Amazon	In Center-West	In South/ Southeast	In Northeast	All except Amazon	In Center-West	In South/ Southeast	In Northeast
Amazon	Small	–4.5	–0.8	–3.7	–2.0	1.4	0.4	–1.0	1.7
	Large	–1.0	0.6	–3.3	–0.8	3.7	0.7	1.4	1.5
	Forest	13.5	2.7	9.2	4.0	8.1	0.8	5.6	4.0
	Subtotal	–1.0	0.1	–1.9	–0.8	3.0	0.6	0.6	2.0
Northeast	Small	–1.2	0.6	–0.3	–2.0	6.6	0.3	–0.7	8.2
	Large	–5.1	0.6	–2.0	–3.2	7.2	0.5	–1.7	9.5
	Forest	0.2	–0.2	1.9	–3.6	–1.8	–0.1	–0.4	–1.8
	Subtotal	–2.5	0.6	–0.8	–2.5	6.5	0.4	–1.0	8.3
Center–West	Small	5.3	–1.8	4.5	2.2	2.3	0.7	0.4	2.0
	Large	–1.5	0.0	–3.6	0.7	3.2	0.4	0.5	2.4
	Forest	–1.0	–1.6	–0.2	–0.7	–0.9	–0.5	–0.7	–0.1
	Subtotal	–0.2	–0.4	–2.0	1.0	3.0	0.4	0.4	2.2
South/ Southeast	Small	1.2	–0.9	2.6	–1.7	2.6	0.2	1.8	0.2
	Large	–4.2	–1.7	–2.5	–3.0	0.5	0.5	–1.1	1.1
	Forest	–3.7	–0.8	–3.6	–0.9	–0.3	0.1	–0.9	0.3
	Subtotal	–1.1	–1.2	0.3	–2.2	1.7	0.3	0.6	0.5
Brazil		–1.2	–0.7	–0.3	–1.7	2.8	0.3	0.3	2.1

(continued)

Table G.1—Continued

		Improvement in livestock				Improvement in annuals, Perennials, and livestock			
Region	Producer	All except Amazon	In Center-West	In South/ Southeast	In Northeast	All except Amazon	In Center-West	In South/ Southeast	In Northeast
Amazon	Small	–16.8	–3.6	–12.9	–4.4	–17.3	–3.1	–15.2	–5.5
	Large	–28.7	–7.2	–18.3	–9.0	–27.6	–5.7	–20.0	–5.9
	Forest	–12.4	–5.1	–9.7	–4.6	–18.2	–6.3	–12.2	–6.3
	Subtotal	–19.9	–4.9	–14.2	–5.9	–20.7	–4.4	–16.3	–5.7
Northeast	Small	–4.9	–2.8	–11.4	13.5	0.1	–1.9	–12.7	17.7
	Large	–13.1	–5.0	–14.2	2.9	–10.7	–4.2	–20.2	6.6
	Forest	1.3	0.5	2.0	–0.9	2.0	1.0	6.8	–5.8
	Subtotal	–7.5	–3.4	–11.9	9.3	–3.5	–2.6	–14.6	13.0
CenterWest	Small	–3.8	17.8	–13.9	–2.7	3.4	20.8	–9.3	0.6
	Large	–11.0	9.5	–13.6	–6.6	–11.0	9.2	–18.0	–4.2
	Forest	0.8	–2.2	2.4	0.6	0.7	–3.8	4.0	0.8
	Subtotal	–9.4	10.9	–13.4	–5.7	–8.1	11.2	–15.9	–3.2
South Southeast	Small	–3.3	–2.4	1.7	–4.4	1.0	–3.2	7.7	–5.5
	Large	–9.3	–4.2	–2.3	–4.3	–14.8	–5.3	–5.3	–6.6
	Forest	–1.3	0.0	–0.4	–0.2	–1.7	0.3	–1.9	0.9
	Subtotal	–5.5	–3.0	0.0	–4.1	–5.3	–3.9	2.2	–5.6
Brazil		–7.7	–1.7	–4.8	–2.2	–6.7	–2.0	–4.4	–2.2

Note: Columns represent type of technological innovation and where it occurs.

References

Abbink, G. A., M. C. Braber, and S. I. Cohen. 1995. A SAM–CGE demonstration model for Indonesia: Static and dynamic specifications and experiments. *International Economic Journal* 9(3): 15–33.

Acevedo, M. F.; D. L. Urban, and M. Ablan. 1995. Transition and gap models of forest dynamics. *Ecological Applications* 5(4): 1040–1055.

Almeida, O. T., and C. Uhl. 1995. Developing a quantitative framework for sustainable resource-use planning in the Brazilian Amazon. *World Development* 23(10): 1745–1764.

Alston, L. J., G. D. Libecap, and B. Mueller. 1999. Theory and applications. A model of rural conflict: Violence and land reform policy in Brazil. *Environment and Development Economics* 4(2): 135–160.

Alston, L. J., G. D. Libecap, and R. Schneider. 1996. *The determinants and impact of property rights: Land titles on the Brazilian frontier.* National Bureau of Economic Research (NBER) Working Paper Series No. 5405. Cambridge, Mass.: NBER.

Alves, D. S. 2002. An analysis of the geographical patterns of deforestation in Brazilian Amazonia in the 1991–1996 period. In *Land use and deforestation in the Amazon,* eds. C. H. Wood and R. Porro, Gainesville, Florida: University Press of Florida.

Alves, D. 2001. O processo de desmatamento na Amazônia. *Parcerias Estratégicas*, Vol.12: 259–275.

Andersen, L. E. 1996. The causes of deforestation in the Brazilian Amazon. *Journal of Environment & Development* 5(3): 309–328.

Andersson, K. 1987. *Taxation of capital in Sweden-A general equilibrium model.* Ph.D. diss., University of Lund, Lund, Sweden.

Arima, E.Y., and C. Uhl. 1997. Ranching in the Brazilian Amazon in a national context: Economics, policy, and practice. *Society and Natural Resources* 10(5): 433–451.

Auerbach, A., and L. Kotlikoff. 1983. National savings, economic welfare, and the structure of taxation. In *Behavioral simulation methods in tax policy analysis*, ed. M. Feldstein. Chicago: University of Chicago Press.

Aylward, B. A. 1993. *The economic value of pharmaceutical prospecting and its role in biodiversity conservation.* Discussion Paper 93–05. London: London Environmental Economics Centre, International Institute for Environment and Development.

Azis, I. J. 1997. Impacts of economic reform on rural–urban welfare: A general equilibrium framework. *Review of Urban and Regional Development Studies* 9(1): 1–19.

Baker, W. L. 1989. A review of models of landscape change. *Landscape Ecology* 2(2): 111–133.

Ballard, C. 1983. Evaluation of the consumption tax with dynamic general equilibrium models. Ph.D. diss., Stanford University, Stanford, Calif., USA.

Ballard, C., and L. Goulder. 1985. Consumption taxes, foresight, and welfare: A computational general equilibrium analysis. In *New developments in applied general equilibrium analysis*, ed. J. Piggott and J. Whalley. Cambridge, UK: Cambridge University Press.

Barbier, E. B. 2001. The economics of tropical deforestation and land use: An introduction to the Special Issue. *Land Economics* 77(2): 155–171.

Bell, C. L. G., and T. N. Srinivasan. 1984. On the uses and abuses of economy-wide models in development policy analysis. In *Economic structure and performance*, ed. M. Syrquin, L. Taylor, and L. E. Westphal. New York: Academic.

Benjamin, N. 1996. Adjustment and income distribution in an agricultural economy: A general equilibrium analysis of Cameroon. *World Development* 24(6): 1003–1013.

Bergman, L. 1988. Energy policy modeling: A survey of general equilibrium approaches. *Journal of Policy Modeling* 10(3): 377–400.

———. 1991. General equilibrium effects of environmental policy: A CGE modeling approach. *Environmental and Resource Economics* 1(1): 43–61.

Bhattacharyya, S. C. 1996. Applied genereal equilibrium models for energy studies: A survey. *Energy Economics* 18(3): 145–164.

Binswanger, H.P. 1991. Brazilian policies that encourage deforestation in the Amazon. *World Development* 19(7): 821–829.

Bourguignon, F., W. Branson, and J. de Melo. 1992. Adjustment and income distribution: A micro-macro model for counterfactual analysis. *Journal of Development Economics* 38: 17–39.

Bovenberg, L. 1984. Capital accumulation and capital immobility: Q-theory in a dynamic general equilibrium framework. Ph.D. diss., University of California, Berkeley, Berkeley, Calif.

———. 1985. Dynamic general equilibrium tax models with adjustment costs. In *Economic equilibrium: Model formulation and solution. Mathematical programming study,* ed. A. Manne. Amsterdam: North-Holland.

Brazil, Ministry of Agrarian Development. 1999. The white book of illegal land appropriation in Brazil. <http://www.desenvolvimentoagrario.gov.br/espaco/pubs/pubs.htm> (accessed September 9, 2002).

Brazil, Ministry of Labor. 2000. Relação annual de informações sociais (RAIS) (on-line database) http://www.mtb.gov.br/Temas/RAIS/Estatisticas/Contendo.RaisOnLine.asp.

Brito, F., and T. Merrick. 1974. Migração, absorção de mão-de-obra e distribução da renda. *Estudos Econômicos* 4: 75–122.

Browder, J. O. 1988. Public policy and deforestation in the Brazilian Amazon. In *Public policies and the misuse of forest resources,* ed. R. Repetto and U. Gillis. Boston: Cambridge University Press.

Brown, D. K. 1992. The impact of a North American free trade area: Applied general equilibrium models. In *North American free trade: Assessing the impact,* ed. N. Lustig, B. P. Bosworth, and R. Z. Lawrence. Washington, D.C.: The Brookings Institution.

Burnham, B. O. 1973. Markov intertemporal land use simulation model. *Southern Journal of Agricultural Economics* 5: 25–38.

Carpentier, C. L., S. A. Vosti, and J. Witcover. Forthcoming. BrasilBEM: A household-farm bioeconomic model for the western Brazilian Amazon forest margin. Environment and Production Technology Division Discussion Paper. International Food Policy Research Institute, Washington, D.C.

Carraro, C., M. Galeotti, and M. Gallo. 1996. Environmental taxation and unemployment: Some evidence on the "double dividend hypothesis" in Europe. *Journal of Public Economics* 62(1): 141–182.

Carvalho, G. O., A. C. Barros, P. Moutinho, and D. Nepstad. 2001. Letter to the editor: Sensitive development could protect Amazonia instead of destroying it. *Nature* 409: 131.

Cattaneo, A. 2001. Deforestation in the Brazilian Amazon: Comparing the impacts of macroeconomic shocks, land tenure, and technological change. *Land Economics* 77(2): 219–240.

Chomitz, K. M., and T. Thomas. 2000. Geographic patterns of land use and land intensity in the Brazilian Amazon. Working Paper. Development Research Group, the World Bank, Washington, D.C.

Convênio INCRA/CRUB/UnB (Instituto Nacional de Colonização e Reforma Agraria, Conselho de Reitores das Universidades Brasileiras, and Universidade de Brasilia). 1997. Censo da reforma agrária no Brasil (summary of a meeting). *Estudos avançados* 11(31): 7–36.

Cottle, R. W., and G. B. Dantzig. 1970. A generalization of the linear complementarity problem. *Journal of Combinatorial Theory* 8 (1): 79–90.

Cottle, R.W., J. Pang, and R. E. Stone. 1992. *The linear complementarity problem.* Boston : Academic Press.

Coxhead, I., and S. Jayasuriya. 1994. Technical change in agriculture and land degradation in developing countries: A general equilibrium analysis. *Land Economics* 70 (1): 20–37.

Coxhead, I., and G. Shively. 1995. *Measuring the environmental impacts of economic changes: The case of land degradation in Philippine agriculture.* University of Wisconsin-Madison, Staff Paper Series No. 384. Madison, Wisc.: University of Wisconsin-Madison.

Coxhead, I., and P. Warr. 1991. Technical change, land quality, and income distribution: A general equilibrium analysis. *American Journal of Agricultural Economics* 73(2): 345–360.

Cruz, W., and R. Repetto. 1992. *The environmental effects of stabilization and structural adjustment programs: The Philippine case.* Washington, D.C.: World Resources Institute.

Darwin, R., M. E. Tsigas, J. Lewandrowski, and A. Raneses. 1995. *World agriculture and climate change: Economic adaptations.* USDA Agricultural Economic Report No. 703. Washington, D.C.: U.S. Department of Agriculture.

Deacon, R. 1994. Deforestation and the rule of law in a cross-section of countries. *Land Economics* 70(4): 414–430.

____. 1999. Deforestation and ownership: Evidence from historical accounts and contemporary data. *Land Economics* 75(3): 341–359.

Deaton, A., and J. Muellbauer. 1980. An almost ideal demand system. *American Economic Review* 70(3): 312–326.

Dee, P.S. 1991. *Modeling steady state forestry in computable general equilibrium context.* Working Paper No. 91/8. Canberra: National Centre for Development Studies.

Dervis, K., J. de Melo, and S. Robinson. 1982. *General equilibrium models for development policy.* Cambridge: Cambridge University Press.

Devarajan, S. 1988. Natural resources and taxation in computable general equilibrium models of developing countries. *Journal of Policy Modeling* 10(4): 505–528.

Devarajan, S., and D. S. Go. 1998. The simplest dynamic general-equilibrium model of an open economy. *Journal of Policy Modeling* 20(6): 677–714.

Devarajan, S., J. D. Lewis, and S. Robinson. 1994. Getting the model right: The general equilibrium approach to adjustment policy. World Bank and International Food Policy Research Institute, Washington, D.C. Photocopy.

Dewatripont, M., and G. Michel. 1987. On closure rules, homogeneity, and dynamics in applied general equilibrium models. *Journal of Development Economics* 26(1): 65–76.

Diao, X., E. Yeldan, and T. L. Roe. 1998. A simple dynamic applied general equilibrium model of a small open economy: Transitional dynamics and trade policy. *Journal of Economic Development* 23(1): 77–101.

Dirkse, S. P., and M. C. Ferris. 1995. The PATH solver: A non-monotone stabilization scheme for mixed complementarity problems. *Optimization Methods and Software* 5: 319–345.

Dixon, P. B., and B. R. Parmenter. 1996. Computable general equilibrium modelling for policy analysis and forecasting. In *Handbook of computational economics,* ed. H. M. Amman, D.A. Kendrick, and J. Rust. Handbooks in Economics 13. Amsterdam: Elsevier/North-Holland.

Dixon, P. B., B. R. Parmenter, J. Sutton, and D. P. Vincent. 1982. *ORANI: A multisectoral model for the Australian economy.* Amsterdam: North-Holland.

Dorfman, R., P. A. Samuelson, and R. M. Solow. 1958. *Linear programming and economic analysis.* New York: McGraw-Hill.

Duarte, R. 1979. Migration and urban poverty in Northeast Brazil. Ph.D. diss., University of Chicago, Chicago, Ill.

Ellerman, A. D., H. D. Jacoby, and A. Decaux. 1998. *The effects on developing countries of the Kyoto Protocol and carbon dioxide emissions trading.* Policy Research Working Paper No.2019. Washington, D.C.: World Bank.

Erlich, S., V. Ginsburgh, and L. Van der Heyden. 1987. Where do real wage decreases lead Belgium? *European Economic Review* 31(7): 1369–1383.

ESALQ (Escola Superior de Agricultura "Luiz de Queiroz"). 1998. *Sistema de informações de fretes para cargas agrícolas* (SIFRECA). Soja (9/97). Piracicaba, Brazil: ESALQ.

Evenson, R. E., and A. F. Avila. 1995. Productivity change and technology transfer in the Brazilian grain sector. *Revista de Economia e Sociologia Rural* 34(2): 93–109.

Faminow, M. D., and S. A. Vosti. 1998. Livestock-deforestation links: Policy issues in the Western Brazilian Amazon. In *Livestock and the environment international conference,* ed. A. J. Nell. Wageningen, the Netherlands: World Bank, Food and Agriculture Organization of the United Nations, and the International Agricultural Centre.

Faminow, M. D., C. Pinho de Sa, and S. J. de Magalhães Oliveira. 1996. Development of an investment model for the smallholder cattle sector in the Western Amazon. International Food Policy Research Institute, Washington, D.C. Photocopy.

Fearnside, P. 1997. Greenhouse gases from deforestation in Brazilian Amazonia: Net committed emissions. *Climatic Change* 35 (3): 321–360.

———. 1999. Forests and global warming mitigation in Brazil: Opportunities in the Brazilian forest sector for responses to global warming under the "clean development mechanism." *Biomass and Bioenergy* 16(3): 171–189.

Ferris, M. C., and C. Kanzow. 1998. Complementarity and related problems: A survey. Mathematical Programming Technical Report 98–17. University of Wisconsin, Madison. Madison, Wisc.

Ferris, M. C., and J. S. Pang. 1997. Engineering and economic applications of complementarity problems. *SIAM Revue* 39(4): 669–713.

FGV (Fundação Getulio Vargas). 1998. Preços de terra (ARIES on-line database: www.fgv.br/cgi-win/aries.exe). Rio de Janeiro.

Findlay, R. 1995. *Factor proportions, trade, and growth.* The Ohlin Lectures No. 5. Cambridge, Mass.: Massachusetts Institute of Technology Press.

Flavin, C. 1989. *Slowing global warming: A worldwide strategy.* Worldwatch Paper 91. Washington, D.C.: Worldwatch Institute.

FAO (Food and Agriculture Organization of the United Nations). 1999. *State of the world's forests.* Rome.

Gasques, J. C., and J. C. da Conceição. 1998. A demanda de terra para a reforma agrária no Brasil. Paper presented at the workshop on Reforma Agrária e Desenvolvimento Sustentável, 23–25 November, 1998, Fortaleza, Brazil.

———. 2000. *Transformações estruturais da agricultura e produtividade total dos fatores.* Discussion Paper No. 768. Rio de Janeiro: Instituto de Pesquisa Econômica Aplicada.

Go, D. S. 1995. External shocks, adjustment policies and investment in a developing economy: Illustrations from a forward-looking CGE model of the Philippines. *Journal of Development Economics* 44(2): 229–261.

Golan, A., G. Judge, and D. Miller. 1996. *Maximum entropy econometrics: Robust estimation with limited data.* New York : Wiley.

Golan, A., G. Judge, and S. Robinson. 1994. Recovering information from incomplete or partial multisectoral economic data. *Review of Economics and Statistics* 76(3): 541–549.

Goldin, I., and G. C. Rezende. 1990. *Agriculture and economic crisis: Lessons from Brazil.* Organisation of Economic Co-operation and Development (OECD) Development Centre Studies. Paris: OECD.

Goldin, I., O. Knudsen, and D. van der Mensbrugghe. 1993. *Trade liberalization: Global economic implications.* Paris and Washington, D.C.: Organization for Economic Cooperation and Development and the World Bank.

Goulder, L., and L. Summers. 1987. *Tax policy, asset prices, and growth: A general equilibrium analysis, NBER Working Paper No. 2128.* Cambridge, Mass.: National Bureau of Economic Research (NBER).

Graham, D. H., and S. Buarque de Holanda. 1971. *Migration, regional and urban growth, and development in Brazil.* São Paulo: University of São Paulo, Instituto de Pesquisas Econômicas.

Harris, J. R., and M. P. Todaro. 1970. Migration, unemployment, and development: A two-sector analysis. *American Economic Review* 60: 126–142.

Hasenkamp, G. 1976. *Specification and estimation of multiple-output production functions.* Lecture Notes in Economics and Mathematical Systems No. 120. Berlin: Springer-Verlag.

Hecht, S. B. 1993. The logic of livestock and deforestation in Amazonia. *BioScience* 43(10): 687–695.

Hertel, T. W. 1990. General equilibrium analysis of U.S. agriculture: What does it contribute? *Journal of Agricultural Economics Resources* 42(3): 3–9.

———. 1997. *Global trade analysis: Modeling and applications.* New York: Cambridge University Press.

Hertel, T. W., and M. E. Tsigas. 1988. Tax policy and U.S. agriculture: A general equilibrium approach. *American Journal of Agricultural Economics* 70(2): 289–302.

Homma, A. K. O., R. T. Walker, F. N. Scatena, A. J. de Conto, R. de Amorim Carvalho, C. A. Palheta Ferreira, and A. Itayguara Moreira dos Santos. 1998. Redução dos desmatamentos na Amazonia: Politica agricola ou ambiental. In *Amazonia: Meio ambiente e desenvolvimento agricola,* ed. A. K. O. Homma. Brasilia: Embrapa.

Horn, H. S. 1975. Markovian processes of forest succession. In *Ecology and evolution of communities,* ed. M. L. Cody and J. M. Diamond. Cambridge, Mass.: Belknap Press.

Hudson, E. A., and D. W. Jorgenson. 1974. U.S. energy policy and economic growth, 1975–2000. *Bell Journal of Economics and Management Sciences* 5(2): 461–514.

IBGE (Instituto Brasileiro de Geografia e Estatistica). 1997a. Matriz de insumo-produto Brasil 1995. IBGE. Rio de Janeiro.

———. 1997b. *Sistema de contas nacionais Brasil 1990–1995/96.* IBGE, Rio de Janeiro.

———. 1997c. *Pesquisa nacional por amostra de domicílios 1996.* IBGE, Rio de Janeiro.

———. 1997d. *Pesquisa de orçamentos familiares 1995/1996.* IBGE, Rio de Janeiro.

———. 1998a. *Censo agropecuário 1995/1996.* IBGE, Rio de Janeiro.

———. 1998b. *Contagem populacional de 1996.* IBGE, Rio de Janeiro.

IMF (International Monetary Fund). 2000. *International Financial Statistics.* Washington, D.C.

INPE (Instituto Nacional de Pesquisas Espaciais). 2002. Monitoring the Brazilian Amazon forest by satellite: 2000–2001. São José dos Campos, Brazil.

IPCC (Intergovernmental Panel on Climate Change). 1996. *Climate change 1995: The science of climate change. Contribution of Working Group I to the Second Assessment Report of the Intergovernmental Panel on Climate Change,* ed. Houghton, et al. Cambridge: Cambridge University Press.

Isard, W., and P. Kaniss. 1973. The 1973 Nobel prize for economic science. *Science* 182: 568–591.

Isard, W., I. J. Azis, M. P. Drennan, R. E. Miller, S. Saltzman, and E. Thorbecke. 1998. *Methods of interregional and regional analysis.* Ashgate, UK: Aldershot.

Johansen, L. 1960. *A multisectoral study of economic growth.* Amsterdam: North-Holland.

Jorgenson, D. W., and P. J. Wilcoxen. 1990. Global change, energy prices, and U.S. economic growth. Harvard Institute of Economic Research Discussion Paper No.1511. Harvard University, Cambridge, Mass.

Kaimowitz, D., and A. Angelsen. 1998. *Economic models of tropical deforestation: A review.* Bogor, Indonesia: Center for International Forestry Research.

Kaimowitz, D., N. Byron, and W. Sunderlin. 1998. Public policies to reduce inappropriate tropical deforestation. In *Agriculture and the environment: Perspectives on sustainable rural development,* ed. E. Lutz. Washington, D.C.: World Bank.

Kaimowitz, D. and J. Smith. 1999. Soybean Technology and the Loss of Natural Vegetation in Brazil and Bolivia. In *Technological change in agriculture and tropical deforestation; Agricultural technologies and tropical deforestation*, A. Angelsen and D. Kaimowitz, eds. (New York; London; CABI Publishing), pp. 195–212.

Karsenty, A. 2000. *Economic instruments for tropical forests: The Congo Basin case.* London: International Institute for Environment and Development.

Keuschnigg, C., and W. Kohler. 1995. Dynamic effects of tariff liberalization: An intertemporal CGE approach. *Review of International Economics* 3(1): 20–35.

Koopmans, T. 1951. Analysis of production as an efficient combination of activities. In *Activity analysis of production and distribution.* New York: Wiley.

Lele, U., V. Viana, A. Verissimo, S. Vosti, K. Perkins, and S. A. Husain. 2000. *Brazil - Forests in the balance: Challenges of conservation with development.* Evaluation Country Case Study Series. World Bank Operations Evaluation Department. Washington, D.C.: World Bank.

Leontief, W. W. 1941. *The structure of the American economy 1919–1929.* New York: Oxford University Press.

Löfgren, H., and S. Robinson. 1999. Spatial networks in multi-region Computable General Equilibrium models. Trade and Macroeconomics Division Discussion Paper No. 35. International Food Policy Research Institute, Washington, D.C.

Lysy, F. J. 1982. The character of general equilibrium models under alternative closures. World Bank, Washington, D.C. Photocopy.

Mahar, D. 1988. *Government policies and deforestation in Brazil's Amazon region.* Environment Department Working Paper No. 7. Washington, D.C.: World Bank.

Manne, A. S. 1985. On the formulation and solution of economic equilibrium models. In *Economic equilibrium: Model formulation and solution,* ed. A. S. Manne. Mathematical Programming Study No. 23. Amsterdam : North-Holland.

Markandya, A. 1997. Economic instruments: Accelerating the move from concepts to practical application. In *Finance for sustainable development: The road ahead.* Proceedings of the Fourth Group Meeting on Financial Issues of Agenda 21, Santiago, Chile, 1997. New York: United Nations.

Martin, R., and S. van Wijnbergen. 1986. Shadow prices and the inter-temporal aspects of remittances and oil revenues in Egypt. In *Natural resources and the macroeconomy,* ed. J. P. Neary and S. van Wijnbergen. Cambridge, Mass.: MIT Press.

Martine, G. 1990. Brazil. In *International handbook on internal migration,* ed. C. B. Nam, W. J. Serow, and D. Sly. New York: Greenwood Press.

Mattos, M. M., and C. Uhl. 1994. Economic and ecological perspectives on ranching in the Eastern Amazon. *World Development* 22(2): 145–158.

Mendelsohn, R. 1994. Property rights and tropical deforestation. *Oxford Economics Papers* 46: 750–756.

Mercenier, J., and T. N. Srinivasan, ed. 1994. *Applied general equilibrium and economic development: Present achievements and future trends.* Ann Arbor, Mich.: The University of Michigan Press.

Moura Costa, P., and C. Wilson. 2000. An equivalence factor between CO2 avoided emissions and sequestration—Description and applications in forestry. *Mitigation and Adaptation Strategies for Global Change* 5(1): 51–60.

Moura Costa, P., J. Salmi, M. Simula, and C. Wilson. 1999. *Financial mechanisms for sustainable forestry.* Report prepared for the United Nations Development Program/Forest Program by INDUFOR/EcoSecurities Limited, London and Helsinki.

Mukherjee, N. 1996. Water and land in South Africa: Economy-wide impacts of reform—A case study for the Olifants River. Trade and Macroeconomics Division Discussion Paper, TMD-DP 12. International Food Policy Research Institute, Washington, D.C.

Nascimento, C.N.B. and A.K.O. Homma. 1984. The Brazilian Enterprise for Agricultural Research, Center for Agroforestry Research of the Eastern Amazon. Belém, Pará, Brazil.

Najberg, S., F. Rigolon, and S. Vieira. 1995. *Modelo de equilibrio geral computavel como instrumento de politica economica: uma analise de cambio e tarifas.* Discussion Paper 30. Rio de Janeiro: Banco Nacional de Desenvolvimento Economico e Social.

Nepstad, D. C., A. Verissimo, A. Alencar, C. Nobre, E. Lima, P. Lefebvre, P. Schlesinger, C. Potter, P. Moutinho, E. Mendoza, M. Cochrane, and V. Brooks. 1999. Large-scale impoverishment of Amazonian forests by logging and fire. *Nature* 398(6726): 505–507.

OECD (Organization for Economic Co-operation and Development). 1990. Modeling the effects of agricultural policies. *OECD Economic Review* (Special Issue 13).

Pearce. D., B. Day, J. Newcombe, T. Brunello, and T. Bello. 1998. The clean development mechanism: Benefits of the CDM for developing countries. Final draft produced by the Center for Social and Economic Research on the Global Environment (CSERGE) for the UK Department for International Development. CSERGE, University College, London.

Pereira, A. M. M. 1988. Corporate tax integration in the United States: A dynamic general equilibrium analysis. PhD. diss., Stanford University, Palo Alto, Calif.

Pereira, A. M., and J. B. Shoven. 1988. Survey of dynamic applied general equilibrium models for tax policy evaluation. *Journal of Policy Modeling* 10 (3): 401–436.

Perlman, J. 1977. *O mito da marginalidade*. Rio de Janeiro: Editora Paz e Terra.

Persson, A. 1995. A dynamic CGE model of deforestation in Costa Rica. In *Biodiversity conservation,* ed. C. A. Perrings et al. Dordrecht, the Netherlands: Kluwer Academic Publishers.

Persson, A., and M. Munasinghe. 1995. Natural resource management and economywide policies in Costa Rica: A computable general equilibrium (CGE) modeling approach. *The World Bank Economic Review* 9(2): 259–285.

Perz, S. G. 2000. The rural exodus in the context of economic crisis, globalization, and reform in Brazil. *The International Migration Review* 34(3): 842–881.

Pfaff, A. S. 1997. *What drives deforestation in the Brazilian Amazon? Evidence from satellite and socioeconomic data.* Policy Research Working Paper No. 1772. Washington, D.C.: World Bank.

Pyatt, G., and J. I. Round. 1985. *Social accounting matrices: A basis for planning.* Washington, D.C.: World Bank.

Reis, E., and S. Margulis. 1991. Options for slowing Amazon jungle clearing. In *Global warming: Economic policy responses,* ed. R. Dornbusch and J. M. Poterba. Cambridge, Mass.: MIT Press.

Reis, E. P., and S. Schwartzman. 1978. Spatial dislocation and social identity building: Brazil. *International Social Sciences Journal* 30(1): 98–115.

Richards, M. 1999. *Internalising the externalities' of tropical forestry: A review of innovative financing and incentive mechanisms.* European Union Tropical Forestry Paper No. 1, London: Overseas Development Institute.

———. 2000. Can sustainable tropical forestry be made profitable? The potential and limitations of innovative incentive mechanisms. *World Development* 28(6): 1001–1016.

Robidoux, B., M. Smart, J. Lester, and L. Bearsejour. 1989. The agriculture expanded GET model: Overview of model structure. Canada, Department of Finance, Ottawa, Canada. Photocopy.

Robinson, S. 1989. Multisectoral models. In *Handbook of Development Economics,* ed. H. Chenery and T. N. Srinivasan. Amsterdam: North-Holland.

———. 1990. Analyzing agricultural trade liberalization with single-country computable general equilibrium models. Department of Agricultural and Resource Economics Working Paper 524. University of California at Berkeley, Berkeley, Calif.

———. 1991. Macroeconomics, financial variables, and computable general equilibrium models. *World Development* 19(11): 1509–1525.

Robinson, S, S. Hoffmann, and S. Subramanian. 1994. Defense spending reductions and the California economy: A computable general equilibrium model. Department of Agricultural and Resource Economics Working Paper 688. University of California at Berkeley, Berkeley, Calif.

Robinson, S., M. Kilkenny, and K. Hanson. 1990. *The USDA/ERS computable general equilibrium (CGE) model of the United States.* ERS Staff Report No. AGES-9049, Agricultural and Rural Economy Division, Economic Research Service. Washington, D.C.: U.S. Department of Agriculture.

Sahota G. 1968. An economic analysis of internal migration in Brazil. *Journal of Political Economy* 76(2): 218–245.

Samuelson, P. 1949. *Market mechanisms and maximization.* Rand Paper Series P-69. Santa Monica, Calif.: Rand Corporation.

Schneider, R. 1992. *Brazil: An analysis of environmental problems in the Amazon.* Report No. 9104 BR, Vol. I. Washington, D.C.: World Bank.

———. 1994. *Government and the economy on the Amazon frontier.* Latin America and the Caribbean Technical Department, Regional Studies Program Report 34. Washington, D.C.: World Bank.

Scholz, C. M. 1998. Involuntary unemployment and environmental policy: The double dividend hypothesis: A comment. *The Scandinavian Journal of Economics* 100(3): 663–664.

SENAR/FGV (Serviço Nacional de Aprendizagem Rural/Fundação Getúlio Vargas) 1998. *O perfil da agricultura Brasileira.* Brasília.

Seroa de Motta, R. 1997. The economics of biodiversity in Brazil: The case of forest conservation. In *Investing in biological diversity: The Cairns conference.* Proceedings of the OECD International Conference on Incentive Measures for the Conservation and Sustainable Use of Biological Diversity, held in Cairns, Australia, 25–28 March, 1996. Paris: Organisation for Economic Co-operation and Development.

Serrão, E. A. S., and A. K. O. Homma. 1993. Country profiles: Brazil. In *Sustainable agriculture and the environment in the humid tropics.* Washington, D.C.: National Academy Press for the National Research Council.

Shoven, J. B., and J. Whalley. 1984. Applied general-equilibrium models of taxation and international trade. *Journal of Economic Literature* 22(3): 1007–1051.

_____. 1992. *Applying general equilibrium.* Cambridge, UK: Cambridge University Press.

Shugart, H. H., T. R. Crow, and J. M. Hett. 1973. Forest succession models: A rationale and methodology for modeling forest succession over large regions. *Forest Science* 19: 203–212.

Simpson, R. D., R. A. Sedjo, J. W. Reid. 1996. Valuing biodiversity for use in pharmaceutical research. *Journal of Political Economy* 104(1): 163–185.

Southgate, D. 1998. *Tropical forest conservation: An economic assessment of the alternatives in* Latin America. New York: Oxford University Press.

Srinivasan, T. N. 1982. General equilibrium theory, project evaluation, and economic development. In *The theory and experience of economic development: Essay in honor of Sir W. Arthur Lewis,* ed. M. Gersowvitz et al. London: George Allen and Unwin.

Stone, S. W. 1998. Evolution of the timber industry along an aging frontier: The case of Paragominas (1990–95). *World Development* 26(3): 433–448.

Summers, L. 1985. Taxation and the size and composition of the capital stock: An asset price approach. Harvard University, Cambridge, Mass. Photocopy.

Terkla, D. 1984. The efficiency value of effluent tax revenues. *Journal of Environmental Economics and Management* 11(2): 107–123.

Thiele, R. 1994. How to manage tropical forests more sustainably: The case of Indonesia. *Intereconomics* 29 (July/August): 184–193.

Thiele, R., and M. Wiebelt. 1992. *Modeling deforestation in a computable general equilibrium model.* Kieler Arbeitspapiere No. 555. Kiel, Germany: Kiel Institute of World Economics.

Toniolo, A., and C. Uhl. 1995. Economic and ecological perspectives on agriculture in the eastern Amazon. *World Development* 23(6): 959–973.

Valentim, J. F., and S. A. Vosti. Forthcoming. *ASB research in the western Brazilian Amazon:* Issues, activities, and impacts. Alternatives to Slash-and-Burn Programme. Nairobi: International Centre for Research in Agroforestry (ICRAF).

Verissimo, A., P. Barreto, M. Mattos, R. Tarifa, and C. Uhl. 1992. Logging impacts and prospects for sustainable forest management in an old Amazonian frontier: The case of Paragominas. *Forest Ecology and Management* 55: 169–199.

Vosti, S. A., C. L. Carpentier, and J. Witcover. 2002. *Agricultural intensification by smallholders in the western Brazilian Amazon: From deforestation to sustainable land use.* Research Report 130. Washington, D.C.: International Food Policy Research Institute.

Vosti, S. A., C. L. Carpentier, J. Witcover, and J. F. Valentim. 2001. Intensified small-scale livestock systems in the western Brazilian Amazon. In *Agricultural technologies and tropical deforestation,* ed. A. Angelsen and D. Kaimowitz. Wallingford, U.K.: CAB International.

Watson, R. T. 2000. *Land use, land-use change, and forestry: A special report of the IPCC.* Cambridge: Cambridge University Press.

Weinhold, D. 1999. Estimating the los of agricultural productivity in the Amazon. *Ecological Economics* 31: 63–76.

White, D., F. Holmann, S. Fujisaka, K. Reategui, and C. Lascano. 2001. Will intensifying pasture management in Latin America protect forests—or is it the other way round? In *Agriculutral technologies and tropical deforestation.* A. Angelsen and D. Kaimowitz, eds. Wallingford, U.K.: CAB International.

Wiebelt, M. 1994. *Protecting Brazil's tropical forests: A CGE analysis of macroeconomic, sectoral and regional policies.* Kieler Arbeitspapiere No. 638. Kiel, Germany: Kiel Institute of World Economics.

Wood, C.H. and R. Porro, eds. 2002. *Land use and deforestation in the Amazon.* Gainesville, Florida: University Press of Florida.

Xie, J., J. R. Vincent, and T. Panayotou. 1996. *Computable general equilibrium models and the analysis of policy spillover in the forest sector.* Environment Discussion Paper No. 19. Cambridge, Mass.: Harvard Institute for International Development.